Learning Tableau

Leverage the power of Tableau 9.0 to design rich data visualizations and build fully interactive dashboards

Joshua N. Milligan

PUBLISHING

BIRMINGHAM - MUMBAI

Learning Tableau

First published: April 2015

Production reference: 1150415

Published by Packt Publishing Ltd.
Livery Place
35 Livery Street
Birmingham B3 2PB, UK.

ISBN 978-1-78439-116-4

www.packtpub.com

Credits

Author
Joshua N. Milligan

Reviewers
David Baldwin
Sagar Kapoor
Joshua Kennedy
Shawn Wallwork

Commissioning Editor
Sarah Crofton

Acquisition Editor
Sonali Vernekar

Content Development Editor
Ritika Singh

Technical Editor
Mrunal M. Chavan

Copy Editors
Deepa Nambiar
Vikrant Phadke
Rashmi Sawant
Stuti Srivastava

Project Coordinator
Judie Jose

Proofreaders
Stephen Copestake
Paul Hindle

Indexer
Monica Ajmera Mehta

Graphics
Sheetal Aute
Abhinash Sahu

Production Coordinator
Nilesh R. Mohite

Cover Work
Nilesh R. Mohite

About the Author

Joshua N. Milligan has been a consultant with Teknion Data Solutions since 2004, where he currently serves as a team leader and project manager. With a strong background in software development and custom .NET solutions, he uses a blend of analytical and creative thinking in BI solutions, data visualization, and data storytelling. His years of consulting have given him hands-on experience in all aspects of the BI development cycle, including data modeling, ETL, enterprise deployment, data visualization, and dashboard design. He has worked with clients in numerous industries, including financial, healthcare, marketing, and government.

In 2014, Joshua was named a Tableau Zen Master, the highest recognition of excellence from Tableau Software. As a Tableau-accredited trainer, mentor, and leader in the online Tableau community, he is passionate about helping others gain insights into their data. He was a technical reviewer of *Tableau Data Visualization Cookbook, Packt Publishing*, and is currently reviewing *Creating Data Stories with Tableau Public, Packt Publishing*. His work has appeared multiple times on Tableau Public's *Viz of the Day* and Tableau's website. Joshua also shares frequent Tableau tips, tricks, and advice on his blog at www.VizPainter.com.

I owe a debt of gratitude to all those who have mentored, guided, and taught me throughout the years. My father, Stuart, opened up the world of computer programming for me when he showed me how I could use code to make the family computer do anything I could imagine. Thank you to all the individuals at Teknion Data Solutions: my colleagues, with whom I have had the privilege to collaborate on a daily basis, and the management and owners, who have made an investment in our training and growth and created an exciting place to build a career. I would also like to thank Tableau employees and members of the online community for creating an incredible place to mutually learn, share, help others, and have fun. I also owe much to the technical reviewers of this book for investing their time and expertise. Most of all, thanks to my wonderful wife, Kara, who has supported and encouraged me all along the way.

About the Reviewers

David Baldwin has provided consulting in the business intelligence sector for 15 years. His experience includes Tableau training and consulting, developing BI solutions, technical writing, project management, and web and graphic design. His vertical experience includes the financial, healthcare, human resource, aerospace, energy, waste management, and entertainment industries. As a Tableau trainer and consultant, David enjoys serving a variety of clients throughout USA. Tableau provides him with a platform that collates his broad experience into a skill set that can service a diverse client base.

Sagar Kapoor is passionate about work in the field of analytics and understanding business processes. He is a team player, with mesmerizing goals and ideas that can change the future course of what is possible in the field of analytics and mobility. He is looking forward to working in different industrial sectors with regard to analytics and providing them with the best results for decision making.

Sagar is currently working with NttData as a business analyst on the client side for Essar Steel India. He leads the implementation of Tableau for his clients across the organization.

Joshua Kennedy is a young business intelligence analyst currently residing in southern California. After studying software design and production in school, he went on to work as a business intelligence consultant. It was during his tenure as a consultant that he gained in-depth knowledge of Tableau and the data business. An accredited Tableau trainer, he has spent the last few years traveling across the USA, working with a variety of industries.

Focusing his skills on helping others learn Tableau, Joshua has a great understanding of new user learning processes with new innovative software. He has published a paper on the benefits of simple business software design principles. He continues to build on his thousands of hours of study and training experience in and out of the business space.

I'd like to thank Joshua Milligan, who offered a reviewer's spot to me and served as my mentor in the years when I was just getting started with Tableau and business intelligence. It has been a pleasure working alongside such a wonderful individual and helping him complete this work.

Shawn Wallwork won the 2014 Tableau Zen Master award for his extensive work on the Tableau forums, answering a variety of questions. He has spent more than 3 years using Tableau almost exclusively to help his clients solve their data analysis and visualization challenges. He works as an independent consultant for clients all over the world, all from his home in Placitas, New Mexico, USA.

www.PacktPub.com

Support files, eBooks, discount offers, and more

For support files and downloads related to your book, please visit www.PacktPub.com.

Did you know that Packt offers eBook versions of every book published, with PDF and ePub files available? You can upgrade to the eBook version at www.PacktPub.com and as a print book customer, you are entitled to a discount on the eBook copy. Get in touch with us at service@packtpub.com for more details.

At www.PacktPub.com, you can also read a collection of free technical articles, sign up for a range of free newsletters and receive exclusive discounts and offers on Packt books and eBooks.

https://www2.packtpub.com/books/subscription/packtlib

Do you need instant solutions to your IT questions? PacktLib is Packt's online digital book library. Here, you can search, access, and read Packt's entire library of books.

Why subscribe?

- Fully searchable across every book published by Packt
- Copy and paste, print, and bookmark content
- On demand and accessible via a web browser

Free access for Packt account holders

If you have an account with Packt at www.PacktPub.com, you can use this to access PacktLib today and view 9 entirely free books. Simply use your login credentials for immediate access.

Table of Contents

Preface

The Tableau community is full of individuals passionate about the software. We use software every day—web browsers, word processors, e-mail applications, instant messaging, and numerous other apps. What is it about Tableau that inspires people to write books and blogs and spend hours volunteering to help others visualize their data?

Tableau is unique in several ways. It is easy and transparent. You can immediately connect to nearly any data source and start asking and answering questions about your data in a visual way. It's also intuitive. Its interface allows hands-on interaction with data, it's easy to get into a flow, and every action uncovers new insights. It's fun! It allows creativity and gives freedom. You're not locked into chart types and wizards that give only one path to a solution. Tableau designers feel like artists, with data as paint and Tableau as a blank canvas.

At the same time, Tableau introduces a paradigm vastly different from traditional BI tools. This book presents the fundamentals for understanding and working within that paradigm. It will equip you with the foundational concepts that will help you use Tableau to explore, analyze, visualize, and share the stories contained in your data.

What this book covers

Chapter 1, Creating Your First Visualizations and Dashboard, introduces the basic concepts of data visualization and multiple examples of individual visualizations, which are ultimately put together in an interactive dashboard.

Chapter 2, Working with Data in Tableau, shows that Tableau has a very distinctive paradigm for working with data. This chapter explores that paradigm and gives examples of connecting to and working with various data sources.

Chapter 3, Moving from Foundational to Advanced Visualizations, expands upon the basic concepts of data visualization to show you how standard visualization types can be extended.

Chapter 4, Using Row-level and Aggregate Calculations, introduces the concepts of calculated fields and the practical use of calculations, and walks through the foundational concepts for creating row-level and aggregate calculations.

Chapter 5, Table Calculations, proves that table calculations are one of the most complex and powerful features in Tableau. This chapter breaks down the basics of scope, direction, partitioning, and addressing to help you understand and use these to solve practical problems.

Chapter 6, Formatting a Visualization to Look Great and Work Well, shows how formatting can make a standard visualization look great, have appeal, and communicate well. This chapter introduces and explains the concept of formatting in Tableau.

Chapter 7, Telling a Data Story with Dashboards, dives into the details of building dashboards and telling stories with data. It covers the types of dashboards, objectives of dashboards, and concepts such as actions and filters. All of this is done in the context of practical examples.

Chapter 8, Adding Value to Analysis – Trends, Distributions, and Forecasting, explores the analytical capabilities of Tableau and demonstrates how to use trend lines, distributions, and forecasting to dive deeper into the analysis of your data.

Chapter 9, Making Data Work for You, explains that data in the real world isn't always structured well. This chapter examines the structures that work best and the techniques that can be used to address data that can't be fixed.

Chapter 10, Advanced Techniques, Tips, and Tricks, builds upon the concepts covered in the previous chapters. This chapter expands your horizons by introducing numerous advanced techniques while giving practical advice and tips.

Chapter 11, Sharing Your Data Story, throws light on the fact that that, once you've built your visualizations and dashboards, you'll want to share them. This chapter explores numerous ways of sharing your stories with others.

What you need for this book

You will need a licensed or trial version of Tableau Desktop to follow the examples contained in this book. You may download Tableau Desktop from Tableau Software at www.tableau.com. Tableau Public is also available as a free download from Tableau and may be used with many of the examples. The examples in this book use the interface and features of Tableau 9.0.

The concepts will apply to other versions, though some interface steps and terminology may vary. The provided workbooks may be opened in Tableau 9.0 or later versions, though you can use any version to connect to the provided data files to work through the examples.

Who this book is for

Anyone seeking to understand their data and enhance their skills to visually explore, analyze, and present their data story to others will greatly benefit from this book. While it is assumed that you have some knowledge of data, you do not need to have in-depth knowledge of databases, SQL scripts, or coding.

This book starts with the foundational principles and builds upon them to acclimate you to advanced concepts. The goal is to give not a series of steps to memorize but a solid understanding of working in the Tableau paradigm. Whether you are just a beginner or have years of experience, this book will further you in the journey of learning—and even mastering—Tableau.

Conventions

In this book, you will find a number of styles of text that distinguish between different kinds of information. Here are some examples of these styles, and an explanation of their meaning.

Code words in text, database table names, folder names, filenames, file extensions, pathnames, dummy URLs, and user input are shown as follows: "We'll create a calculated field named Floor to determine whether an apartment is upstairs or downstairs."

A block of code is set as follows:

```
IF [Apartment] >= 1 AND [Apartment] <= 3
 THEN "Downstairs"
ELSEIF [Apartment] > 3 AND [Apartment] <= 6
 THEN "Upstairs"
ELSE "Unknown"
END
```

New terms and **important words** are shown in bold. Words that you see on the screen, for example, in menus or dialog boxes, appear in the text like this: "Drag and drop the **Customer** field onto the **Rows** shelf."

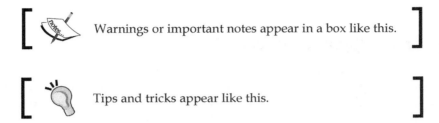

Warnings or important notes appear in a box like this.

Tips and tricks appear like this.

Reader feedback

Feedback from our readers is always welcome. Let us know what you think about this book—what you liked or may have disliked. Reader feedback is important for us to develop titles that you really get the most out of.

To send us general feedback, simply send an e-mail to feedback@packtpub.com, and mention the book title via the subject of your message.

If there is a topic that you have expertise in and you are interested in either writing or contributing to a book, see our author guide on www.packtpub.com/authors.

Customer support

Now that you are the proud owner of a Packt book, we have a number of things to help you to get the most from your purchase.

Downloading the example code

You can download the example code files for all Packt books you have purchased from your account at http://www.packtpub.com. If you purchased this book elsewhere, you can visit http://www.packtpub.com/support and register to have the files e-mailed directly to you.

Downloading the color images of this book

We also provide you a PDF file that has color images of the screenshots/diagrams used in this book. The color images will help you better understand the changes in the output. You can download this file from: https://www.packtpub.com/sites/default/files/downloads/LearningTableau_GraphicsBundle.pdf.

Errata

Although we have taken every care to ensure the accuracy of our content, mistakes do happen. If you find a mistake in one of our books—maybe a mistake in the text or the code—we would be grateful if you would report this to us. By doing so, you can save other readers from frustration and help us improve subsequent versions of this book. If you find any errata, please report them by visiting http://www.packtpub.com/submit-errata, selecting your book, clicking on the **errata submission form** link, and entering the details of your errata. Once your errata are verified, your submission will be accepted and the errata will be uploaded on our website, or added to any list of existing errata, under the Errata section of that title. Any existing errata can be viewed by selecting your title from http://www.packtpub.com/support.

Piracy

Piracy of copyright material on the Internet is an ongoing problem across all media. At Packt, we take the protection of our copyright and licenses very seriously. If you come across any illegal copies of our works, in any form, on the Internet, please provide us with the location address or website name immediately so that we can pursue a remedy.

Please contact us at copyright@packtpub.com with a link to the suspected pirated material.

We appreciate your help in protecting our authors, and our ability to bring you valuable content.

Questions

You can contact us at questions@packtpub.com if you are having a problem with any aspect of the book, and we will do our best to address it.

1
Creating Your First Visualizations and Dashboard

Tableau is an amazing data visualization platform! With it, you will be able to achieve incredible data discovery, data analysis, and data storytelling. You will accomplish all of these tasks and goals visually. In fact, Tableau is unique among all other data visualization tools because it uses **VizQL**, a visual query language. This means you won't write a lot of tedious SQL or MDX or painstakingly work through wizards to select a chart type and then link components together with data.

Instead, you will be interacting with the data in a visual environment and Tableau will automatically translate your actions into the necessary queries behind the scenes. Much of your work will be drag and drop. Tableau empowers you to work with data rapidly and iteratively, switch visualization types on-the-fly, and ask new questions and gain new insight.

This chapter introduces the foundational principals of data visualization in Tableau. You will take on the role of an analyst for a coffee chain. We'll work through a series of examples that will introduce the basics of connecting to data, exploring and analyzing the data visually, and finally, putting it all together in a fully interactive dashboard. These concepts will be developed far more extensively in subsequent chapters. This chapter lays the foundation, including:

- Connecting to data in **Access**
- Creating bar charts
- Creating line charts
- Creating geographic visualizations
- Using **Show Me**
- Putting everything together in a dashboard

Connecting to data in Access

Tableau connects to data stored in a wide variety of files and databases. This includes flat files such as Excel and text files, relational databases such as SQL Server and Oracle, cloud-based data sources such as Google Analytics and Amazon Redshift, and OLAP data sources such as Microsoft SQL Server Analysis Services. With very few exceptions, the process of building visualizations and performing analysis will be the same no matter what data source you use. We'll cover the details of connecting to different data sources in later chapters.

For now, we will connect to an **Access** data source included in the resources supplied with this book. This chapter's workbook includes a connection to the data source, but we will walk through the steps of connecting using a new workbook first:

1. Open Tableau. The home screen should appear. If you are not on the home screen, then from the menu, navigate to **Data | New Data Source**.

2. Under **Connect** in the **To a file** section, click on **Microsoft Access**.

3. Click on **Browse…** and navigate to the `Learning Tableau\Data` directory and open `Coffee Chain.mdb`.

4. The database does not contain any security, so you do not need to adjust either the **Database Password** or **Workgroup Security** options. Click on **OK** to connect.

5. Tableau's data connection screen allows you to visually create connections to data sources. We'll look at the details of this screen in the next chapter. For now, drag the **CoffeeChain Query** table, located on the left-hand side under **Table**, into the center of the screen. This query contains all the fields we'll need in order to build our initial visualizations and the dashboard. The query functions look like a single table in the Tableau connection.

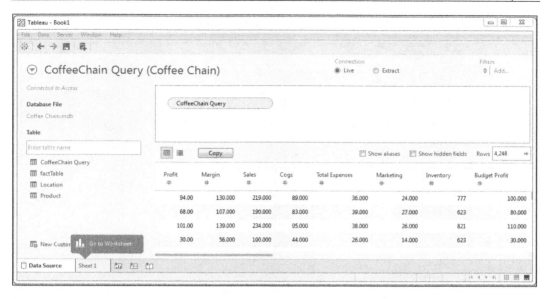

6. Once you have configured a data connection, you can create visualizations and dashboards by clicking on a tab at the bottom. Click on the **Sheet 1** tab in the lower-left corner.

7. If the **Show Me** panel is displayed in the upper-right corner, collapse it by clicking at the top of the panel where the **Show Me** text appears.

You should now be in the main work area of Tableau, which looks like this:

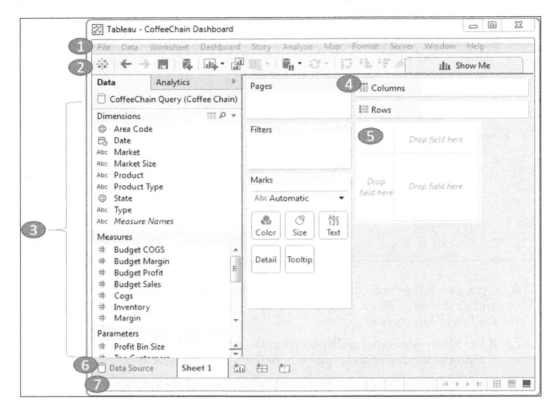

Locate the numbered features of the main workspace. We'll refer to these features throughout the book, so familiarize yourself with the terminology:

- **1**: The menu allows you to perform a wide variety of functions.

- **2**: The toolbar makes common functions, such as undo, redo, save, connect to data, and so on, easily available.

- **3**: The sidebar appears on the left-hand side and contains different features and controls based on your current task (for example, data visualization, applying analytics, formatting, or designing dashboards). The default sidebar in the main workspace consists of two tabs allowing you easy access to data and analytics. The **Data** window consists of data source connection(s), and the fields contained in the data sources are divided into **Dimensions** and **Measures**.

- **4**: You can drag and drop data fields from the **Data** window onto various shelves, such as **Pages, Filters, Columns, Rows, Color, Size,** and **Text**.

- **5**: The canvas (sometimes called the view or visualization) is where Tableau will draw visualizations based on where you drag and drop fields. You may also drag and drop fields directly onto the canvas.

 Observe the distinction between fields that are in the **Data** window and fields that are in the view. Fields listed in the **Data** window are available to add to a view. Fields that you have dropped onto shelves or the canvas are **in the view** or **active fields**.

- **6**: The tabs give you easy access to the data connection screen and also to each sheet, dashboard, and story you create in the current workbook. At the bottom is a single tab named **Sheet 1**. As you build your workbook, you will most likely add new sheets (often also called views), which are individual visualizations. You may also add dashboards (these combine and relate multiple sheets and other components together on a single screen) and even stories (these combine multiple dashboards and views into a unified data story). A set of buttons next to the existing tabs allows the quick creation of new sheets, dashboards, or stories.

- **7**: The status bar will give you information about the current view and includes some options on the right-hand side to navigate sheets in the workbook.

Having created your connection to the data, you are ready to begin to visualize and analyze the data. Over the course of the following examples, you'll take on the role of an analyst for the coffee chain. We'll build multiple visualizations that answer various questions and finally put everything together in an interactive dashboard. As we begin, let's consider a few of the foundational principles.

Foundations for building visualizations

When you first connect to a data source such as the coffee chain data, Tableau will display the data connection and the fields in the **Data** window in the sidebar on the left-hand side. Fields can be dragged from the **Data** window onto various parts of the view on the right-hand side. Fields can be dropped onto the canvas area or onto various shelves, such as **Rows, Columns, Color,** or **Size**. The placement of the fields will result in different encodings of the data based on the type of field.

The fields from the data source are visible in the **Data** window and are divided into **Measures** and **Dimensions**. Understanding the difference between **Measures** and **Dimensions** is essential.

> **Measures** are values that are aggregated. That is, they can be summed, averaged, counted, or they can have a minimum or maximum.
>
> **Dimensions** are values that determine the level of detail at which measures are aggregated. You can think of them as slicing the measures or creating groups into which the measures fit. The combination of dimensions used in the view defines the view's basic level of detail.

As an example, consider a view created using the **Sales** and **Market** fields from the coffee chain connection.

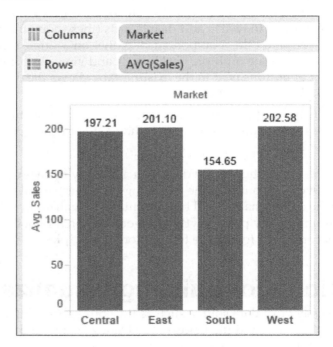

The **Sales** field is used as a measure in this view. Specifically, it is aggregated as an average. When you use a field as a measure in the view, the type aggregation (for example, **SUM**, **MIN**, and **MAX**) will be shown on the active field. Note in the preceding example that the active field on **Rows** clearly indicates the average aggregation of sales: **AVG(Sales)**.

The **Market** field is a dimension with one of four values for each record of data: **Central**, **East**, **South**, or **West**. When the field is used as a dimension in the view, it slices the measure. So instead of an overall average, the view in the preceding example shows you the average sales for each market.

Tableau makes it easy to recategorize fields and change default aggregations.

You can recategorize a field in the **Data** window as a dimension or measure by simply dragging the field from **Measures** to **Dimensions** or vice versa.

You can recategorize a field in the view as a dimension or measure by right-clicking on the field in the view and then selecting **Dimensions** or **Measures**.

You can change the default type of the aggregation of a measure by right-clicking on a **Measures** field in the **Data** window and navigating to **Default Properties | Aggregation**.

You can change the type of aggregation of a field in the view by right-clicking on the field in the view, selecting **Measures**, and then selecting the desired type of aggregation.

Another important distinction to make with fields is whether a field is being used as discrete or continuous. Tableau will give you a visual indication of the default for a field (the color of the icon in the **Data** window) and how it is being used in the view (the color of the active field on a shelf). Discrete fields are blue; continuous fields are green.

Whether a field is discrete or continuous, it determines how Tableau visualizes it based on where it is used in the view.

Discrete (blue) fields have values that are shown as distinct and separate from each other. Discrete values can be reordered and still make sense.

When a discrete field is used on the **Rows** or **Columns** shelves, the field defines row or column headers.

Market			
Central	East	South	West

When used for color, a discrete field defines a discrete color palette in which each color describes a distinct value of the field.

Market   ■ Central   ■ East   ■ South   ■ West

Continuous (green) fields have values that are shown as flowing from one field to another. Numeric and date fields are often used as continuous fields in the view. The values of these fields have an order that it would make no sense to change.

When used on **Rows** or **Columns**, a continuous field defines an axis.

$0	$2,000,000	$4,000,000	$6,000,000	$8,000,000
		Sales		

When used for colors, a continuous field defines a gradient.

Sales   24,031 [████████████████] 128,311

Most dimensions are discrete by default, and most measures are continuous by default. However, any numeric or date field, whether dimension or measure, can be used as a continuous field in the view. Any field, whether dimension or measure, can be used as a discrete field in the view.

To change the default of a field, right-click on the field in the **Data** window and select **Convert to Discrete** or **Convert to Continuous**. To change how a field is used in the view, right-click on the field in the view and select **Discrete** or **Continuous**.

As you work through the examples in this chapter, pay attention to the fields you are using to create the visualizations, whether they are dimensions or measures, and whether they are discrete or continuous. Experiment with changing fields in the view from continuous to discrete and vice versa to gain an understanding of the difference in the visualization.

Visualizing data

A new connection to a data source is an invitation to explore. At times, you may come to the data with very well-defined questions and a strong sense of what you expect to find. Other times, you will come to the data with general questions and very little idea of what you will find. The data visualization capabilities of Tableau empower you to rapidly and iteratively explore the data, ask new questions, and make new discoveries.

The visualization examples in the following sections cover a few of the most foundational chart types. As you work through the examples, keep in mind that the goal is not simply to learn how to create each chart type. Rather, the examples are designed to help you think through the process of discovery, analysis, and storytelling. Tableau is designed to make this process intuitive, rapid, and transparent. Far more important than memorizing steps to create a bar chart is to understand how to use Tableau to create a bar chart and then iteratively adjust your visualization to gain new insight as you ask new questions.

Bar charts

Bar charts visually represent data in a way that makes comparisons of a value across different categories easy to compare. Length is the primary means by which you will visualize the data. You may also incorporate color, size, stacking, and order to communicate additional attributes and values.

Creating bar charts in Tableau is very easy. Simply drag and drop the measure you want to see on either the **Rows** or **Columns** shelf and the dimension that defines the categories onto the opposing **Rows** or **Columns** shelf.

As an analyst for the coffee chain, you are ready to begin a discovery process focused on profit. Create a new sheet for this and every subsequent example in the workbook you started. This chapter's workbook will contain the complete examples, so you can compare your results.

Begin your analysis with the following steps:

1. Drag and drop the **Profit** field from **Measures** on the **Data** window on the left to the **Columns** shelf. You now have a bar chart with a single bar representing the sum of profit for all the data in the data source.

2. Drag and drop the **Product Type** field from **Dimensions** in the **Data** window to the **Rows** shelf. This slices the data to give you four bars, representing the sum of profit for each product type.

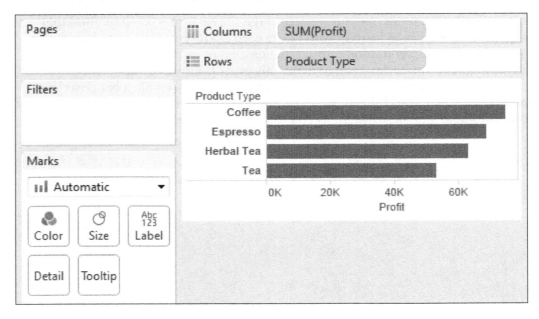

You now have a horizontal bar chart. This makes the comparison of profit between the product types easy. Notice how the mark type dropdown on the **Marks** card is set to **Automatic** and shows that Tableau has determined that bars are the best visualization given the fields you have placed in the view. As a discrete dimension, the **Product Type** field defines row headers for each product type in the data. As a continuous measure, the **Profit** field defines an axis that causes each bar to be drawn from 0 to the value of the total profit for each product type.

Typically, Tableau draws a mark (a bar, shape, circle, square, and so on) for every intersection of dimensional values in the view. In this simple case, Tableau draws a single bar mark for each dimensional value (**Coffee**, **Espresso**, **Herbal Tea**, and **Tea**) of **Product Type**. The type of mark is indicated and can be changed in the drop-down menu on the **Marks** card. The number of marks drawn in the view can be observed on the lower-left status bar.

Tableau draws different marks in different ways. For example, bars are drawn from 0 (or the end of the previous bar, if stacked) along the axis. Circles and other shapes are drawn at locations defined by the value(s) of the field defining the axis. Take a moment to experiment with selecting different mark types from the dropdown on the **Marks** card. Having an understanding of how Tableau draws different mark types will help you master the tool.

Extending bar charts for deeper analysis

Using the preceding bar chart, you can easily see that **Coffee** has more total profit than any other product type and **Tea** has less total profit than any other type. What if you want to further understand the profitability of product types in different markets?

Drag the **Market** field from **Dimensions** in the **Data** window to the left of the **Product Type** field that is already on the **Rows** shelf.

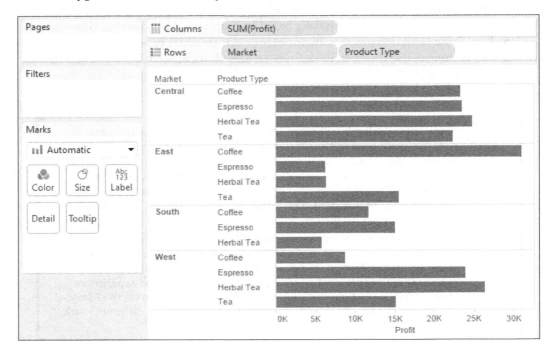

You still have a horizontal bar chart. But now you've introduced **Market** as another dimension that changes the level of detail in the view and further slices the aggregate value of profit. By placing **Market** before **Product Type**, you are able to easily compare sales for each product type within a given market.

Now you are starting to make some discoveries. For example, **Coffee** is making a bit more profit than any other product type in the **East** market but a bit less than other products in the **West** market.

Let's take a look at a different view using the same fields arranged differently.

Drag the **Market** field from the **Rows** shelf and drop it on the **Color** shelf.

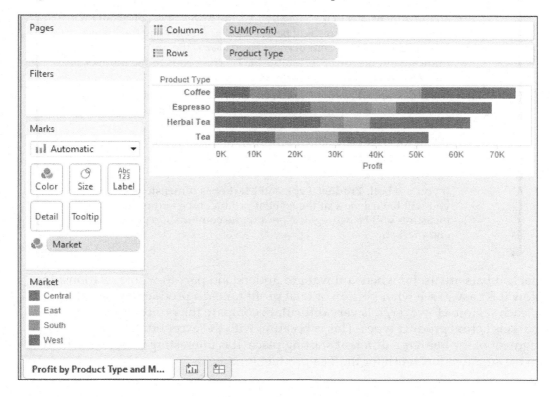

Instead of a side-by-side bar chart, you now have a stacked bar chart. Notice how each segment of the bar is color-encoded by the **Market** field. Additionally, a color legend has been added to the workspace. You haven't changed the level of detail in the view, so the profit is still summed for every combination of market and product type.

The **level of detail** of a view is a key concept when working with Tableau. In most basic visualizations, the combination of values of all the dimensions in the view defines the lowest level of detail for that view. All measures will be aggregated or sliced by the lowest level of detail. In the case of most basic views, the number of marks (indicated in the lower-left status bar) corresponds to the number of intersections of dimensional values.

If, for example, **Product Type** is the only field used as a dimension, you will have a view at the product type level of detail and all measures in the view will be aggregated per product.

If **Market** is the only field used as a dimension, you will have a view at the market level of detail and all measures in the view will be aggregated per market.

If you use both **Product Type** and **Market** as dimensions in the view, you will have a view at the level of product type and market. All measures will be aggregated per unique combination of product type and market.

Stacked bars are useful when you want to understand part-to-whole relationships. Now it is easy to see what portion of total profit for each product type is made in each region. However, it is very difficult to compare the profit for most of the markets across product types. This is because, with the exception of **West**, every segment of the bar has a different starting place. It is interesting to note that there is no bar segment for **South** for the **Tea** product type.

Conclude this example with the following steps:

1. Experiment by dragging the **Market** field from **Color** to the other various shelves on the **Marks** card, such as **Size**, **Label**, and **Detail**. Observe that in each case, the bars remain stacked but are redrawn based on the visual encoding defined by the **Market** field.

2. Finally, drag the **Market** field off the shelf and drop it in a gray area on the workspace. This will remove that field from the view. You should now have a bar chart that looks like the very first one you created. Later, you'll be able to use this simple bar chart in an interactive dashboard.

3. Right-click on the tab labeled **Sheet 1** at the bottom of the screen and rename the sheet Profit by Product Type.

4. From the **File** menu, select **Save** and save your workbook as CoffeeChainAnalsyis.twb.

At the time of writing this, Tableau does not have an autosave feature. You will want to get into the habit of saving the workbook early and then pressing *Ctrl + S* or selecting **Save** from the **File** menu often to avoid losing your work.

Line charts

Line charts connect related marks in a visualization to show movement or the relationship between connected marks. The position of the marks and lines that connect them are the primary means of communicating the data. Additionally, you can use size and color to visually communicate additional information.

The most common kind of line chart is a **time-series**. Time series show the movement of values over time. They are very easy to create in Tableau and require only a date and a measure.

Continue your analysis of coffee chain profit using the workbook you just saved:

1. Create a new worksheet by clicking on the new worksheet button immediately to the right of the **Profit by Product Type** tab at the bottom. You should now have a new, blank worksheet named **Sheet 2** by default. If you accidentally create a new dashboard or story, you can delete it or use the undo button.

2. Rename the sheet to `Profit over Time` by right-clicking on the tab and selecting **Rename** or by double-clicking on the text of the tab.

3. Drag the **Profit** field from **Measures** to **Rows**. This gives you a single, vertical bar representing the sum of all profit in the data source.

4. To turn this into a time series, you must introduce a date. Drag the **Date** field from **Dimensions** in the **Data** window and drop it in **Columns**.

5. Tableau has a built-in date hierarchy and the default level of detail has given you a line chart connecting 2 years. Let's say you want to see the profit by month. Right-click on the **YEAR(Date)** field in **Columns** and select **Month** from the second set of dates. Future chapters will discuss the details of the various date options.

You can right-click drag and drop a date field as a shortcut to selecting the date option available from this menu. Right-click dragging and dropping other types of fields is a shortcut to quickly selecting other customizations, such as the type of aggregation. If you are using Tableau on a Mac, the equivalent is *Option* + left-click + drag and drop.

You now have a time series that shows you the profit for each month.

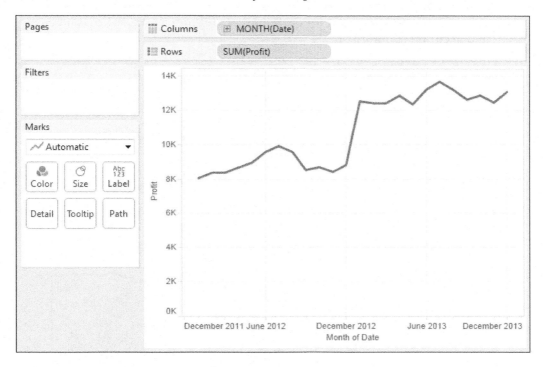

Extending line charts for deeper analysis

Right now, you are looking at the overall profit over time. Let's perform some analysis at a slightly deeper level:

1. Drag the **Market** field from **Dimensions** to **Color**. Now, you have a line per market with each line in a different color and a legend indicating which color is used for which market. As with the bars, adding a dimension to Color splits the marks. However, unlike the bars where the segments were stacked, the lines are not stacked. Instead, the lines are drawn at the exact value for the sum of profit for each market and month. This allows an easy and accurate comparison.

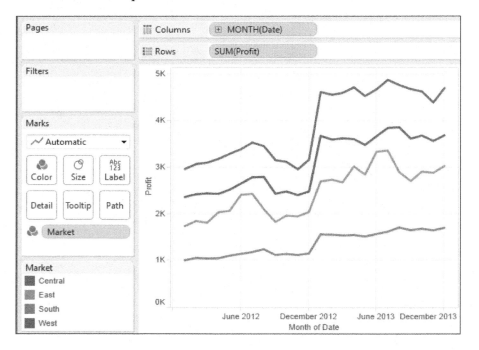

2. Drag the **Product** field from **Dimensions** and drop it directly on top of the **Market** field currently in the **Marks** card. This replaces the **Market** field with **Product**. You now have 13 overlapping lines. Often, you'll want to avoid more than two or three overlapping lines. However, note that you can enable highlighting on the color legend using the icon in the upper-right section of the legend. Then, clicking on a product in the color legend will highlight the associated line in the view. This can be a good way to pick out a single item and compare it with all others.

3. Hold down the *Ctrl* key and then drag the **Product** field from **Color** in the **Marks** card and drop it in **Rows**. This copies the field in the view and you now have a line chart for each product. Now you have a way to compare each product over time without overwhelming the overlap. This is the start of a spark-lines visualization that will be developed more fully when advanced visualizations are discussed.

4. Remove the **Product** field from the **Rows** shelf to return to the first time series created in this exercise. Additionally, you may experiment with the undo button in the toolbar.

Geographic visualizations

Tableau makes creating geographic visualizations very easy. The built-in geographic database allows any field recognized as playing a geographic role to define a latitude and longitude. This means that even if your data does not contain latitudes and longitudes, Tableau will provide them for you based on fields such as **Country**, **State**, **City**, or **Zip Code**. If your data contains **Latitude** and **Longitude** fields, you may use them instead of the generated values.

Although most databases do not strictly define geographic roles for fields, Tableau will automatically assign geographic roles to fields based on the field name and a sampling of values in the data. You can assign or reassign geographic roles to any field by right-clicking on the field in the **Data** window and using the **Geographic Role** option. This is also a good way to see what built-in geographic roles are available.

The power and flexibility of Tableau's geographic capabilities as well as the options for customization will be covered in more detail in *Chapter 10, Advanced Techniques, Tips, and Tricks*. In the following examples, we'll consider some of the foundational concepts of geographic visualizing.

Geographic visualization is incredibly valuable when you need to understand where things happen and if there are any spatial relationships within the data. Tableau offers two basic forms of geographic visualization:

* Filled maps
* Symbol maps

Filled maps

Filled maps, as the name implies, make use of filled areas, such as the country, state, county or zip code, to show the location. The color that fills the area can be used to encode values of measures or dimensions.

What if you want to understand profit for your coffee chain and see whether there are any patterns geographically? Let's take a look at some examples of how you might do this:

1. Create a new sheet and name it `Profit by Location`.

2. Double-click on the **State** field in the **Data** window. Tableau automatically creates a geographic visualization using the **Latitude (generated)**, **Longitude (generated)**, and **State** fields.

3. Drag the **Profit** field from the **Data** window and drop in on the **Color** shelf in the **Marks** card. Based on the fields and shelves you've used, Tableau has switched the automatic mark type to filled maps.

The filled map fills each state with a single color to indicate the relative sum of profit for each state. The color legend, now visible in the view, gives the range of values and indicates that the state with the least profit had a total of $799 and the state with the most profit had a total of $31,785.

You may observe that not all states are shown. Tableau will only draw a geographic mark, such as a filled state, if it exists in the data and is not excluded by a filter. There were only 20 states in the data and therefore, there are only 20 filled states. The rest of the map is a background image.

Filled maps can work well in interactive dashboards and have quite a bit of aesthetic value. However, certain kinds of analyses are very difficult with filled maps. Unlike other visualization types where size can be used to communicate facets of the data, the size of a filled geographic region only relates the geographic size. For example, which state has the highest profit? You might be tempted to say California, but are you sure that's not just because it is larger than Illinois? Which has more profit: Massachusetts or Texas? Use filled maps with caution and consider pairing them with other visualizations for clear communication.

Symbol maps

The other standard type of geographic visualization available in Tableau is a symbol map. Marks on this map are not drawn as filled regions; rather, marks are shapes or symbols placed at specific geographic locations. Size, color, and shape may also be used to encode additional dimensions and measures.

Continue your analysis using the `Profit by Location` sheet you developed previously:

1. Drag **Area Code** from **Dimensions** to the **Detail** shelf on the **Marks** card. Tableau automatically switches to a symbol map and draws a circle at the latitude and longitude of each area code.

2. Drag **Sales** from **Measures** to the **Size** shelf on the **Marks** card. At this point, you have color encoded by profit and size encoded by sales. This allows some very useful analysis, as you can immediately identify areas with high sales and low profit. Profit may not have been as useful on the **Size** shelf because it can have negative values and there is no way to visually represent a negative size.

You can improve upon the default view. Click on the **Color** shelf and set the transparency to somewhere between 50 percent and 75 percent. Additionally, add a dark border. This makes the marks stand out and you can better discern any overlapping marks.

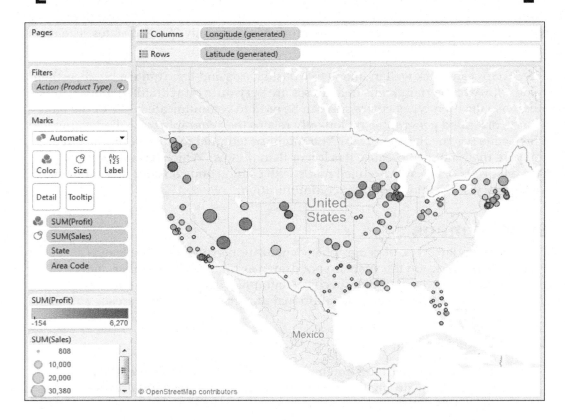

Unlike filled maps, symbol maps allow you to use size to visually encode aspects of the data. Symbol maps also allow greater precision. In fact, if you have latitude and longitude in your data, you can very precisely plot marks at a street address level of detail. This type of visualization also allows you to map locations that do not have clearly defined boundaries. Note that if you were to change the mark type from **Automatic** to **Filled Map** in the view, you would get an error message indicating that filled maps are not supported at the level of detail in the view.

Using Show Me

Show Me is a powerful component of Tableau. The **Show Me** toolbar displays small thumbnail images of different types of visualizations, allowing you create visualizations with a single click. Based on the fields you select in the **Data** window and the fields that are already in the view, **Show Me** will enable possible visualizations and highlight a recommended visualization. Explore the features of **Show Me** by following these steps:

1. Create a new worksheet in the workbook and name it ShowMe Example.

2. If the **Show Me** window is not expanded, click on the **Show Me** button in the upper-right section to expand the window.

3. Press and hold the *Ctrl* key while clicking on the **Area Code**, **State**, and **Profit** fields in the **Data** window to select each of these fields.

Observe that the **Show Me** window has enabled certain visualization types, such as text tables, heat maps, symbol maps, filled maps, and bar charts. These are the visualizations that are possible given the fields already in the view in addition to any fields selected in the data window. **Show Me** also gives a description of what fields are needed for a given visualization type. Symbol maps, for example, require one geographic dimension and up to 2 measures.

Other visualizations are grayed out, such as line charts and histograms. These visualization types cannot be created with the fields that are currently in the view and selected in the **Data** window. Hover over the grayed out line charts in **Show Me**. **Show Me** indicates that line charts require one or more measures, which you have selected, but also require a date field, which you have not selected.

> Tableau will actually draw line charts with fields other than dates. **Show Me** gives you options for what is typically considered good practice for visualizations. Understanding how Tableau renders visualizations based on fields and shelves instead of always relying on **Show Me** will give you much greater flexibility in your visualizations. At the same time, you will need to cultivate an awareness of good visualization practices.

Finally, note that with **Area Code**, **State**, and **Profit** selected, the symbol map has a blue border in the **Show Me** window. **Show Me** indicates the symbol map as most likely the best visualization for the selected fields.

Show Me can be a powerful way to quickly iterate through different visualization types as you search for insight into the data. However, as a data explorer, analyst, and storyteller, you should consider **Show Me** as a helpful guide that gives suggestions. You may know that a certain visualization type will answer your questions more effectively than the suggestions from **Show Me**. You may have a plan for a visualization type that will work well as part of a dashboard but isn't even included in **Show Me**.

You will be well on your way to learning and mastering Tableau when you can use **Show Me** effectively, but feel just as comfortable building visualizations without it. **Show Me** is powerful to quickly iterate through visualizations as you look for insight and raise new questions. It is useful to start with a standard visualization that you will further customize. It is wonderful as a teaching and learning tool.

Be careful not to use it as a crutch without understanding how visualizations are actually built from the data. Take time to evaluate why certain visualizations are or are not possible. Pause to see what fields and shelves were used when you selected a certain visualization type.

Conclude the **Show Me** example by following these steps:

1. Experiment with **Show Me** by clicking on various visualization types, looking for insight into the data that may be more or less obvious with different ways of visualizing the data. Circle views and box-and-whisker-plots show the distribution of area code for each state. Bar charts easily expose several area codes with negative profit.

2. Right-click on the current **Show Me Example** sheet tab at the bottom and select **Delete Sheet**.

Bringing everything together in a dashboard

Often, you'll need more than a single visualization to communicate the full story of the data. In these cases, Tableau makes it very easy for you to use multiple visualizations together on a dashboard. In Tableau, a dashboard is a collection of views, filters, parameters, images, and other objects that work together to communicate a data story. Dashboards are often interactive and allow end users to explore different facets of the data.

Dashboards serve a wide variety of purposes and can be tailored for a wide variety of audiences. Consider the following possible dashboards:

- A summary level view of profit and sales to allow executives to have a quick glimpse into the current status of the company
- An interactive dashboard allowing sales managers to drill into sales territories to identify threats or opportunities
- A dashboard allowing doctors to track patient readmissions, diagnoses, and procedures in order to make better decisions about patient care
- A dashboard allowing the management of a real estate company to identify trends and make decisions for various apartment complexes
- An interactive dashboard for loan officers to make lending decisions based on portfolios broken down by credit ratings and geographic location

Considerations for different audiences and advanced techniques will be covered in great detail in *Chapter 7, Telling a Data Story with Dashboards*. For now, let's consider an example that introduces foundational examples.

Continue with the `CoffeeChainAnalysis` dashboard you have been building, and click on the new dashboard button to the right of the **Profit by Location** tab. You now have a blank dashboard, and the sidebar on the left-hand side shows options to build a dashboard instead of the **Data** window that was visible in a worksheet. The sidebar should look like this:

The dashboard window consists of several key components. Techniques to use these objects will be detailed later. For now, focus on gaining some familiarity with the options that are available.

First is a list of all visible worksheets in the dashboard. You can add these to a dashboard by dragging and dropping them. A light gray shading will indicate the location of the sheet once it is dropped. You can also double-click on any sheet, and it will be added automatically.

The next section lists multiple additional objects that can be added to the dashboard. **Horizontal** and **Vertical** layout containers will give you finer control over the layout; **Text** allows you to add text labels and titles. Images and even embedded web content can be added. Finally, a **Blank** object allows you to preserve blank space in a dashboard or may serve as a placeholder.

Next, you can select whether new objects will be added as **Tiled** or **Floating**. **Tiled** objects will snap into a tiled layout next to other tiled objects or within layout containers. **Floating** objects will float on top of the dashboard in successive layers.

The **Layout** section gives you the ability to navigate through a hierarchical structure of objects that have been added to the dashboard. This can be useful for finding objects in complex dashboards.

The final section gives you control over the size of the dashboard as well as pixel-perfect sizing and positioning of floating objects.

Building your dashboard

Continue following these steps to build the dashboard:

1. Successively double-click on each sheet listed in the **Dashboard** section on the left-hand side: **Profit by Product Type**, **Profit Over Time**, and **Profit by Location**. Notice that double-clicking on an object adds it to the layout of the dashboard.

When a worksheet is first added to a dashboard, any legends, filters, or parameters that were visible in the worksheet view will be added to the dashboard. If you want to add these types of objects, select the sheet in the dashboard and click on the little drop-down caret in the upper-right corner of the sheet and locate the object you want to add. Nearly every object has the drop-down caret providing many options to fine-tune appearance and control its behavior:

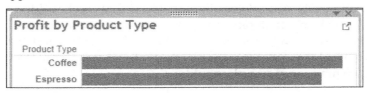

2. Add a title of `Profit Analysis` to the dashboard by dragging the **Text** object from the sidebar to the top of the dashboard, enter the text, and change the size to **24** pt. You may need to resize the text object using the selection outline. Alternately, you can show the default title for the dashboard by checking the **Title** option at the bottom of the left sidebar.

3. Select the **Profit by Product Type** sheet in the dashboard and click on the drop-down caret in the upper-right section. Navigate to **Fit | Entire View**. The **Fit** options describe how the visualization should fill any available space.

 Be careful when using the various **Fit** options. If you are using a dashboard with a size that has not been fixed or if your view dynamically changes the number of items displayed based on interactivity, then what might have looked good once might not fit the view nearly as well.

4. Select the **Sales** size legend by clicking on it. Click on **X** in the upper-right section to remove the legend from the dashboard:

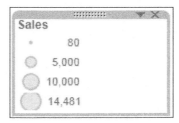

5. Double-click on the title of the **Profit by Location** sheet in the dashboard to edit the text. Add text to indicate that size indicates sales. Having removed the legend, this is the only way the end user will know what size means.

6. Select the **Profit** color legend and use the grab bar at the top of the selection outline to drag and drop the **Profit** color legend immediately below the map.

7. Select the **Profit by Product Type** sheet and click on the drop-down caret in the upper-right section again. Click on **Use as Filter**.

8. Edit the title **Profit by Product Type** and add the `click a bar to see details` text. It is a good practice to give the end user of a dashboard some indication of what interactivity is possible.

9. Click on the bar for **Tea**. The rest of the dashboard should update. Both **Profit by Location** and **Profit over Time** are now filtered by the **Tea** product type:

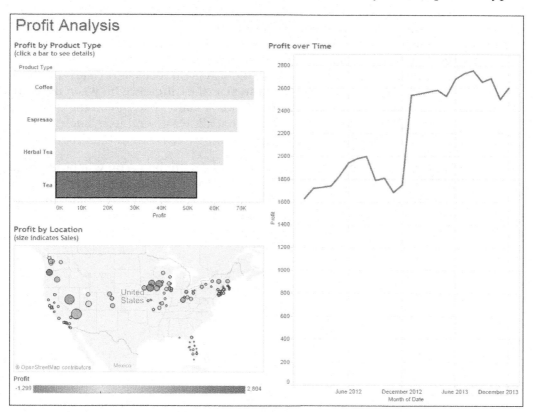

You have created a dashboard that allows interactive analysis. As an analyst for the coffee company, your visualizations allowed you to explore and analyze data. The dashboard you created can be shared with the management as a tool to help them see and understand the data in order to make better decisions. When a manager selects the **Tea** product type, it immediately becomes obvious that there is one location where sales are quite high but the profit is actually a loss. This may lead to decisions such as a change in marketing or removing tea from the inventory at that location. Most likely, it will require additional analysis to determine the best course of action. In this case, Tableau will empower you to continue the iterative process of discovery, analysis, and storytelling.

Summary

Tableau's visual environment allows a rapid and iterative process of exploring and analyzing data visually. You took your first steps in understanding how to use the platform. You connected to an **Access** database and used it to explore and analyze the data using some foundational visualization types, such as bar charts, line charts, and geographic visualizations. Along the way, you focused on learning the techniques and understanding key concepts such as the difference between measures and dimensions and discrete and continuous fields. Finally, you put all the pieces together to create a fully functional dashboard that allows an end user, such as management, to understand your analysis and make discoveries of their own.

In the next chapter, we'll explore how Tableau works with data. You will be exposed to fundamental concepts and practical examples of how to connect to various data sources. Combined with the foundational concepts you just learned about building visualizations, you will be well equipped to move on to more advanced visualizations, deeper analysis, and telling fully interactive data stories.

Working with Data in Tableau

Tableau offers the ability to connect to virtually any data. It does so using a unique paradigm that allows it to leverage the power and efficiency of the existing database engines with an option to extract data locally. This chapter focuses on the foundational concepts of how Tableau works with data, including the following:

- The Tableau paradigm
- Connecting to data
- Working with extracts
- Metadata and sharing connections
- Joins and blends
- Filtering data

The Tableau paradigm

Tableau connects directly to native data engines and also includes the option to extract data locally. The unique experience of working with data in Tableau is a result of VizQL: visual query language.

VizQL was developed as a Stanford research project, focusing on the natural ways humans visually perceive the world and how that could be applied to data visualization. We naturally see differences in size, shape, spatial location, and color. VizQL allows Tableau to translate your actions, as you drag and drop fields of data in a visual environment, into a query language that defines how the data encodes those visual elements. You will never need to read, write, or debug VizQL. As you drag and drop fields onto various shelves defining size, color, shape, and spatial location, Tableau will generate the visual query language behind the scenes. This allows you to focus on visualizing data.

One of the benefits of VizQL is that it provides a common way of describing how the arrangement of various fields in a view defines a query of the data. This common baseline can then be translated into numerous flavors of SQL, MDX, and TQL (**TQL** stands for **Tableau Query Language**, which is used for extracted data). Tableau will automatically perform the translation of VizQL into an optimized query to be run natively by the source data engine.

At its simplest, the Tableau paradigm of working with data looks like the following diagram:

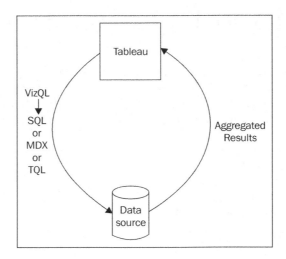

When you drag and drop a field to a shelf in the view, Tableau generates a VizQL query that is then translated into an optimized query for the data source. The query is parsed by the source data engine and results are returned to Tableau, where the visualization engine renders the view based on the resulting data set.

A simple example

Let's say you create a view like this by dropping the **Market** dimension on **Columns** and the **Sales** measure on **Rows**:

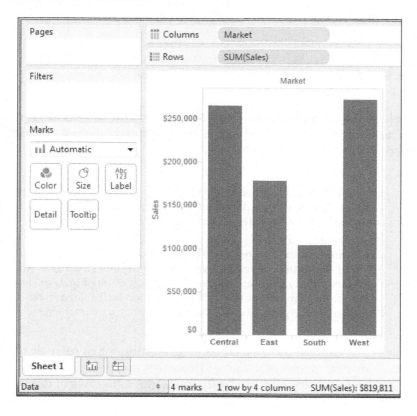

The **Market** field is discrete (blue) and so defines **Columns** headers. As a dimension, it defines the level of detail in the view and slices the measure. The **Sales** field is a measure that is aggregated by summing each sale within each market. As a continuous (green) field, **Sales** defines an axis.

For the purpose of this example, let's say you are connected live to a SQL Server database with the coffee chain data stored in a table called CoffeeChainData. When you first create this view, Tableau generates a VizQL script that is translated into a TSQL script and sent to the SQL Server. The TSQL query would look something like this:

```
SELECT [CoffeeChainData].[Market] AS [none:Market:nk],
 SUM([CoffeeChainData].[Sales]) AS [sum:Sales:ok]
FROM [dbo].[CoffeeChainData] [CoffeeChainData]
GROUP BY [CoffeeChainData].[Market]
```

This script selects the market and the sum of sales from the table and groups by market. The script aliases the field names using a syntax that can be used by Tableau's engine when the data is returned.

You do not need to understand SQL or any kind of scripting to use Tableau. As you design data visualizations, you will most likely not be concerned with what scripts Tableau generates.

On occasions, a database administrator may want to understand what scripts are running against a certain database to debug performance issues or determine more efficient indexing or data structures. Many databases supply profiling utilities or log execution of queries. In addition, you can find SQL or MDX generated by Tableau in the logs located in the \My Tableau Repository\Logs directory.

There may have been hundreds, thousands, or even millions of rows of sales data in SQL Server. However, when SQL Server processes the query it returns aggregate results. In this case, SQL Server returns only four aggregate rows of data—one row for each market.

To see the aggregate data, Tableau used to draw the view, press *Ctrl + A* to select all the bars, then right-click on one of them, and select **View Data**:

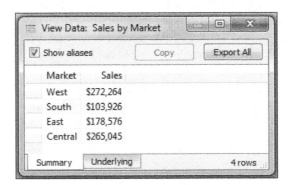

The **Summary** tab displays the aggregate level data that makes up the view. The **Sales** column here is the sum of sales for each market. When you click on the **Underlying** tab, Tableau will query the data source to retrieve all the records that make up the aggregate records. In this case, there are 4,248 underlying records:

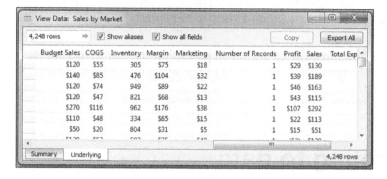

Tableau did not need 4,248 records to draw the view and did not request them from the data source until the **Underlying** data tab was clicked.

Database engines are optimized to perform aggregations on data. Typically, these database engines are also located on powerful servers. Tableau leverages the optimization and power of the underlying data source. In this way, Tableau can visualize massive datasets with relatively little local processing of the data.

Additionally, Tableau will only query the data source when you make changes that require a new query or a view to be refreshed. Otherwise, it will use the aggregate results stored in a local cache, as shown in the following diagram:

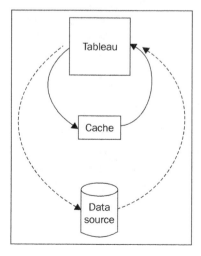

In the preceding example, the query based on the fields in the view (**Market** as a dimension and the sum of **Sales** as a measure) will only be issued once to the data source. When the aggregate results are returned, they are stored in cache. Then, if you were to move **Market** to another visual encoding shelf, such as **Color**, or **Sales** to a different visual encoding shelf, such as **Size**, then Tableau will retrieve the aggregate rows from the cache and simply re-render the view.

 You can force Tableau to bypass the cache and refresh the data from the data source by pressing *F5* or selecting your data source from the **Data** menu and selecting **Refresh**. Do this any time you want a view to reflect the most recent changes in a live data source.

Connecting to data

There is virtually no limit to the data Tableau can visualize. Each successive version of Tableau adds new native connections. Additionally, for any database without a native connection, Tableau gives you the ability to use a generic ODBC connection. With the release of the Extract API, first available in Tableau 8.0, any data source or sources can be programmatically combined and stored in a data extract for use in Tableau.

This section will focus on some practical examples of connecting to various data sources. We won't cover every connection, but will cover several that are representative of others. You may or may not have access to some of the data sources in the following examples. Feel free to follow them if you are able to or merely observe the differences.

You may have multiple data source connections to different sources in the same workbook. Each connection will show up under the **Data** tab on the left-hand side bar.

To add a connection, perform one of the following:

- Select **Connect to Data** from the home screen
- Select the **Data Source** tab on the workspace controls in the lower-left corner
- From the menu, select **Data | New Data Source**

Any of these actions will take you to the **Connect** screen. This interface allows you to select a data source and configure the settings for the connection. A distinction is made between the two main sources of data:

- **In a file**: This means that the data is in a file located in a local or network directory.

- **On a server**: This means that the data is located in a database or service that resides on a server. This server could be the same machine where Tableau Desktop is installed, another physical or virtual machine on the network, or a service in the cloud.

Connecting to data in a file

File-based data includes all sources of data where the data is stored in a file. File-based data sources include the following:

- **Tableau Data Extract**: This is a .tde file containing data that was extracted from an original source. When you connect to a .tde file, the connection retains information about the .tde file but not about the original source.

- **Microsoft Access**: This is a .mdb or .accdb database file created in Access.

- **Microsoft Excel**: This is a .xls, .xlsx, or .xlsm spreadsheet created in Excel.

- **Text File**: This is a delimited text file, most commonly .txt, .csv, or .tab. Multiple text files in a single directory may be joined together in the **Connection** window.

- **Statistical File**: This is an .sav, .sas7bdat, .rda, or .rdata file generated by SAS or R.

- **Other Files**: Select this option to connect to any file data source. In addition to those mentioned, you can connect to Tableau files to import connections you have saved in another tableau workbook (.twb or .twbx). The connection will be imported and changes will only affect the current workbook.

As an example, consider a connection to an Excel file. Here, **Microsoft Excel** has been selected and the `Sample - Superstore Subset (Excel).xlsx` file located at `\My Tableau Repository\Datasources\` was opened, as shown in the following screenshot:

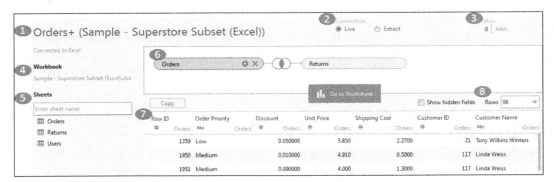

The resulting interface allows you to customize the connection. The following components of the interface are important to note:

- **1**: Click on the Connection Name to change it. Use the dropdown to switch to a different data source.

- **2**: Select the Connection Type, **Live** or use **Extract**.

- **3**: These are data source filters. Click on **Add…** to add a new data source filter. These filters are applied at the data source level and any data excluded will not be available for visualizations.

- **4**: The Source Type text will describe the kind of data source (for example, database file and server). Clicking on the link allows you to select a different source of the same kind or configure initial settings (for example, select a new Excel file or select a different server path).

- **5**: This section lists the different Tables and views available for this connection. For an Excel file, this is a list of sheets and named **ranges**. In a database, it would be a list of tables and views. For text files, it would be a list of all text files in the selected directory. These tables may be joined together in a single connection, as described in the *Joins and blends* section.

- **6**: The Selected Tables workspace allows you to add tables of data, define joins, and configure options. When you hover over a table that has been added, you will see a configuration icon and a remove icon. The configuration icon will give various options for the table based on the data source. For Excel, you may specify whether a given sheet includes field names or not. We'll look at the various join options available later in this chapter.

- **7**: This is for preview, metadata, and options. This section of the interface allows you to preview rows of data that are present in the data source as defined by the configuration, tables, and joins you have specified. Additionally, you can switch to a metadata view that allows you to easily see all the fields in the data, the original tables, and original field names. In the preview, you can see the field types indicated by icons and whether the field will be continuous (green) or discrete (blue) by default. You may use the dropdown from any field to rename or hide the field.

  We'll look at some advanced features and options to clean and restructure data in *Chapter 9, Making Data Work for You*. Additional options give you control over how many rows are displayed in the preview, whether the preview displays aliases, and whether fields you have hidden are visible in the preview.

- **8**: The **Tableau Data Interpreter** can help clean up the data. Excel worksheets may contain multiple layers of headers, footers, and excess information or formatting that should not be included in the data you analyze. When you turn on the data interpreter, it will attempt to filter out the clutter.

- **9**: Once all configuration and setup are complete, click on a sheet tab to begin building visualizations using the data.

If you need to edit the connection at any time, select **Data** from the menu, locate your connection, and then select **Edit Data Source...**. Alternately, you may right-click on any data source under the **Data** tab on the left-hand sidebar and select **Edit Data Source**; or click on the **Data Source** tab in the lower-left corner.

Prior to version 8.2, Tableau used the Microsoft JET driver to connect to Access, Excel, and text files. Tableau now uses a new connection that not only avoids limitations that were present in the JET driver, but also removes the option for writing custom SQL. If you need to use the legacy connection, you can select that option using the drop-down arrow on the **Open File** dialog. This option is not available for the Mac version.

Connecting to data on a server

Database servers such as SQL Server, MySQL, Vertica, and Oracle host data on one or more server machines and use powerful database engines to store, aggregate, sort, and serve data based on queries from client applications. Tableau can leverage the capabilities of these servers to retrieve data for visualization and analysis. Alternately, data can be extracted from these sources and stored in **Tableau Data Extract (TDE)**.

As an example of connecting to a server data source, consider connecting to SQL Server. As soon as the Microsoft SQL Server connection is selected, the interface displays options for some initial configuration:

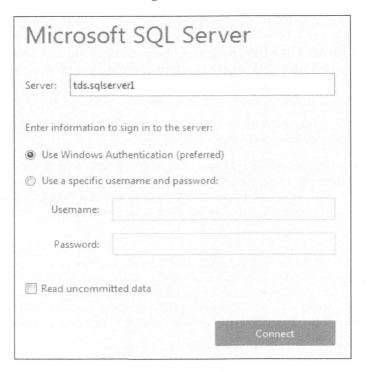

A connection to SQL Server requires the server name as well as authentication information. A database administrator can configure SQL Server to use Windows authentication or a SQL Server username and password. With SQL Server, you can also optionally allow reading of uncommitted data. This can potentially improve performance, but may also lead to unpredictable results if data is being inserted, updated, or deleted at the same time Tableau is querying.

In order to maintain high standards of security, Tableau will not save a password as part of a data source connection. This means that if you share a workbook using a live connection with someone else, they will need to have credentials to access the data. This also means that when you first open the workbook, you will need to re-enter your password for any connections requiring a password.

Once you click on the orange **Connect** button, you will see a screen that is very similar to the connection screen you saw for Excel. The main difference is on the left, where you have an option to select a database:

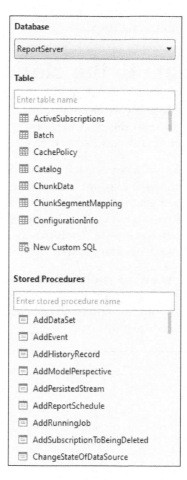

Once you've selected a database, you may add any of the following to the selected tables area:

- **Table**: This represents data tables and views contained in the selected database.

- **New Custom SQL**: You may write your own custom SQL scripts and add them as tables. You may join these as you would any other table or view.

- **Stored Procedures**: You may use a stored procedure that returns a table of data. You will be given the option of setting values for stored procedure parameters or using or creating a Tableau parameter to pass values.

Once you have finished configuring the connection, click on the orange **Go to Worksheet** button to begin visualizing the data.

Connecting to data in the cloud

Certain data connections are made to data that is hosted in the cloud. These include Amazon Redshift, Google Analytics, Google BigQuery, Salesforce, and others. In many cases, you will want to read the documentation provided by the service provider to understand the connection and structure of the data.

It is beyond the scope of this book to cover each connection in depth, but as an example of a cloud data source, we'll explore connecting to Google Analytics. Google Analytics allows users to embed a script in a website. This allows Google to record data about website traffic and web page visits. Google provides online dashboards, but Tableau can connect directly to Google Analytics to allow you to build your own custom dashboards and achieve deeper analysis.

Upon clicking on the **Google Analytics** connection on the **Connect to Data** screen, you will be prompted to log in to your Google account and verify that Tableau Desktop should have access to your Google Analytics account. You will then be given options to customize the connection:

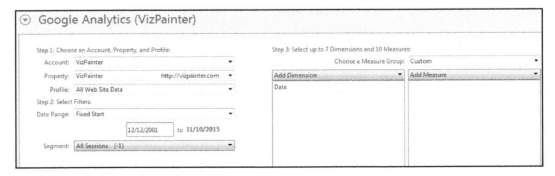

The **Google Analytics** connection dialog allows you to select your account, property, and profile. The **Date Range** and **Segment** fields limit the amount of data pulled from Google that can be useful, as Google will begin to return sampled data when certain row-count thresholds are reached. You are then able to select up to 7 dimensions and 10 measures. Dimensions include things such as the city, state, and country of visitors to a website along with the date, browser, page visited, and much more. Measures include data such as visits, web page views, and visit duration.

> Be aware that some dimensions and measures in Google Analytics are not compatible with each other. Selecting certain dimensions or certain measures may cause others to be disabled from selection. This is expected based on the structure of the data within Google Analytics.

Google Analytics is one of several connections that do not allow for a live connection. Instead, the data is pulled from Google Analytics into a local extract. You can refresh the extract at any time by right-clicking on the data source in the **Data** window, or selecting it from the **Data** menu and selecting **Extract** | **Refresh**. Any changes to the connection properties will cause data to be re-extracted. Extracts are explained in detail in the following sections.

Shortcuts for connecting to data

You can make certain connections very quickly. These options will allow you to begin doing analysis more quickly:

- Paste data from the clipboard. If you have copied data from a spreadsheet, a table on a web page, or a text file, you can often paste the data directly into Tableau. This can be done using *Ctrl + V* or **Data** | **Paste Data** from the menu. The data will be stored as a file and you will be alerted to its location when you save the workbook.

> In some cases, pasting the data into Excel first and then copying from Excel and pasting into Tableau can yield more consistent results. Alternatively, you might also try first pasting to and then copying from a text editor such as Notepad.

- Select **File** | **Open** from the menu. This will allow you to open any data file that Tableau supports such as text files, Excel files, Access files, and even offline **cube (.cub)** files.
- Drag and drop a file from Windows Explorer onto the Tableau workspace. Any valid file-based data source can be dropped onto the Tableau workspace.

- Duplicate an existing connection. You can duplicate an existing data source connection by right-clicking on it and selecting **Duplicate**.

Working with extracts instead of live connections

Most data sources allow the option of either connecting live or extracting the data. However, some cloud-based data sources require an extract. Conversely, OLAP data sources cannot be extracted and require live connections.

When using a live connection, Tableau issues queries directly to the data source (or uses data in the cache if possible). When you extract the data, Tableau pulls some or all of the data from the original source and stores it in a TDE file. Extracts extend the way in which Tableau works with data. Consider the following diagram:

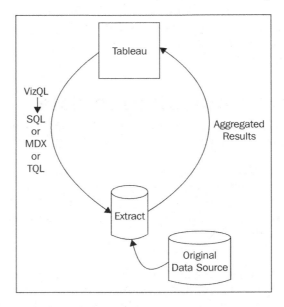

The fundamental paradigm of how Tableau works with data does not change, but you'll notice that Tableau is now querying and getting results from the extract. Data can be retrieved from the source again to refresh the extract. Thus, each extract is a snapshot of the data source at the time of the latest refresh. Extracts offer the benefit of being portable and extremely efficient.

Creating extracts

Extracts can be created in multiple ways, as follows:

- Select **Extract** on the **Connect to Data** screen. The **Edit...** link will allow you to set configuration for the extract:

- Select the data source from the **Data** menu or right-click on the data source on the **Data** window and select **Extract**. You will be given the opportunity to set configuration options for the extract:

- Create an extract using the Tableau Data Extract API. This API allows you to use Python or C/C++ to programmatically create an extract file. The details of this approach are beyond the scope of this book, but relevant documentation is readily available on Tableau's website.

- Certain analytics platforms, such as Alteryx, natively support the creation of Tableau extracts.

When you first create or subsequently configure an extract, you will be prompted to select certain options, as shown here:

You have several options when configuring an extract:

- You may optionally add **Extract Filters** that limit the extract to a subset of the original source. In this example, only data for the East and West markets (except data for the state of New York) will be included in the extract.

- You may aggregate an extract. Data will be rolled up to the level of visible dimensions and optionally to a date level such as year or month.

Visible fields are those that are shown in the **Data** window and are available to be used in a view or calculated fields. You may hide a field from the **Connect to Data** screen or from the **Data** window on the left by right-clicking on a field and selecting **Hide**, as long as it is not used in any view in the workbook or in any calculated field. **Hidden Fields** are not available to be used in a view or in a calculated field. Hidden fields are not included in an extract.

In the preceding example, if only the **Market** and **State** dimensions were visible, the resulting extract would only contain nine rows of data (one row for each state in the East and West markets, excluding the New York state). All **Measures** are aggregated according to their default aggregation.

• You may include all rows or a sampling of the top N rows in the dataset. If you select **All rows**, you can indicate an incremental refresh. If your source data incrementally adds records and you have a field, such as an **Identity** column or **Date** field that can be used reliably to identify new records as they are added, an incremental extract can allow you to add those records to the extract without recreating the entire extract.

Using extracts

Any data source that is using an extract will have a distinctive icon that indicates the data has been pulled from an original source into an extract.

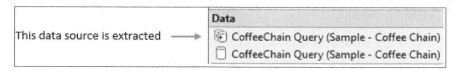

The first data connection in the **Data** window, shown in the preceding screenshot, is extracted while the second is not. After an extract has been created, you may choose to use the extract or not. From the menu, select **Data** and then the data source; alternatively, right-click on the data source in the **Data** window and check or uncheck **Use Extract**. When not using the extract, Tableau will query the original data source.

There are also several other options available from that menu:

- **Extract Data…** will give you options to configure and recreate the extract.

- **Extract | Refresh** will reload data in the extract from the original source but will not rebuild or reoptimize the extract.

- **Extract | Append Data from File** or **Extract | Append Data from Data Source** will allow you to add data, with the same structure, to an existing extract. This can be quite useful when you simply want to append rows. For example, if you have a process where a new Excel file is generated daily, always with the same structure, you can append that data to an extract.

- **Extract | Optimize** will restructure the extract, based on changes you've made since originally creating the extract, to make it as efficient as possible. For example, certain calculated fields may be **materialized** (calculated once so that the resulting value can be stored) and newly hidden columns or deleted calculations will be removed from the extract.

- **Extract | Remove** removes the definition of the extract, optionally deletes the extract file, and resumes a live connection to the original data source.

- **Extract | History** or **Extract | Properties** enables us to view the history and properties of the extract.

Performance

Extracts are very efficient. In fact, extracts are generally faster than most live connections, except for a few extremely efficient columnar databases. This is the result of several factors:

- Extracts are columnar and very efficient to query.

- Extracts are structured, so they can be loaded quickly into memory without additional processing and moved between memory and disk storage. Hence, the size is not limited to the amount of RAM available.

- Many calculated fields are materialized in the extract. The precalculated value stored in the extract can often be read faster than executing the calculation every time the query is executed.

You may choose to use extracts to increase performance over traditional databases. To maximize your performance gain, consider the following guidelines:

- Prior to creating the extract, hide unused fields. If you have created all desired visualizations, you can click on the **Hide Unused Fields** button on the **Extract** dialog to hide all fields not used in any view or calculation.

- If possible, use a subset of data from the original source. For example, if you have historical data for the last 10 years, but will only need the last 2 years for analysis, then filter the extract by the **Date** field.

- Optimize an extract after creating or editing calculated fields, or deleting or hiding fields.

- Store extracts on solid-state disks or drives that are defragmented regularly.

Portability and security

Imagine that you are working with a database on a server on the local network at the office, but you have a long flight that evening and would like to wrap up some analysis on the plane. Normally, you'd have to be onsite to work with the data. With an extract, you can take the data with you.

A .tde file contains all the data extracted from the source. When you save a workbook, you may save it as a **Tableau Workbook (TWB)** file or a **Tableau Packaged Workbook (TWBX)** file. A workbook (.twb) contains definitions for all the connections, fields, visualizations, and dashboards but does not contain any data or external files such as images. When you save a packaged workbook (.twbx), any extracts and external files are packaged together in a single file with the workbook.

A packaged workbook using extracts can be opened with Tableau Desktop or Tableau Reader and published to Tableau Public or Tableau Online.

 A packaged workbook file (.twbx) is really just a compressed ZIP file. If you rename the extension from .twbx to .zip, you can access the content as you would any other ZIP file.

There are a couple of security considerations to be kept in mind when using an extract:

- The extract is made using a single set of credentials. Any security layers that limit what data can be accessed according to the credentials used will not be effective after the extract is created. An extract does not require a username or password. All data in an extract can be read by anyone.

- Any data for visible (nonhidden) fields contained in an extract file (`.tde`) or in an extract contained in a packaged workbook (`.twbx`) can be accessed even if the data is not shown in visualizations. Be very careful when distributing extracts or packaged workbooks containing sensitive or proprietary data.

The story is told of an employee who sent a packaged workbook containing HR data to others in the company. Even though none of the dashboards displayed sensitive data, the extract contained it. It wasn't long before everyone in the company knew everyone else's salary and the original individual was no longer an employee.

When to use an extract

You should consider various factors when determining whether or not to use an extract. In some cases, you won't have an option (for example, OLAP requires a live connection and some cloud-based data sources require an extract). In other cases, you'll want to evaluate the options.

In general, use an extract when:

- You need better performance than you can get with a live connection.
- You need the data to be portable.
- You are using legacy (JET driver) connections to Excel, Access, or text files and you need to use functions such as COUNTD (count distinct) that are not supported by the JET driver but are supported when the data is extracted.
- You want to share a packaged workbook. This is especially true if you want to share a packaged workbook with someone who uses the free Tableau Reader, which can only read packaged workbooks with extracts.

In general, do not use an extract when:

- You have sensitive data. However, you may hide sensitive fields prior to creating the extract, in which case they are no longer part of the extract.

- You need to manage security based on login credentials. (However, if you are using Tableau Server, you may still use extracted connections hosted on Tableau Server that are secured by login credentials. We'll consider sharing your work with Tableau Server in *Chapter 11, Sharing Your Data Story*.)

- You need to see changes in the source data updated in real time.

- The volume of data makes the time required to build the extract impractical. The number of records that can be extracted in a reasonable amount of time will depend on factors such as the data types of fields, the number of fields, the speed of the data source, and network bandwidth.

Metadata and sharing data source connections

Metadata refers to information about the data itself. Nearly every database contains some metadata. This includes lists of the tables and fields for a data source and what type of data (for example, numeric, string, or date) a field contains. Tableau provides an additional layer of metadata that makes it very easy to customize a data connection to add or change attributes of the data.

Customizing a data source

Right-clicking on a field in the **Data** window reveals a menu of metadata options. Some of these options will be demonstrated in the following exercise, and others will be explained in later chapters. These are some of the options available via right-click:

- Rename the field

- Redefine the field as a dimension or a measure

- Change the default use of a field to either discrete or continuous (this can only be done for date and numeric fields)

- Add or remove the field from a hierarchy

- Change the data type of the field

- Change the geographic role of a field

- Change defaults for how a field is displayed in a visualization such as the default colors and shapes, number or date format, sort order (for dimensions), or type of aggregation (for measures)

- Change aliases for values of a dimension

 Metadata options that relate to the visual display of the field such as the default sort order or default number format define the overall default for a field. However, these default settings can be overridden in any individual worksheet by right-clicking on the active field on the shelf and selecting the desired options.

In addition to the preceding options, Tableau's metadata includes additional features such as calculated fields, sets, and whether the fields in the **Data** window are grouped by the database table or user-defined folders.

Sharing a data source

None of the options mentioned change anything in the source database. Instead, the metadata is stored as part of the Tableau workbook. Additionally, you can save a data source connection, with all the metadata, as a **Tableau Data Source (TDS)** file. To do this, right-click on the data source connection or select it from the **Data** menu and select **Add to Saved Data Sources…**.

A TDS (.tds) file is not much more than a copy of the section in the workbook that defines the data source connection:

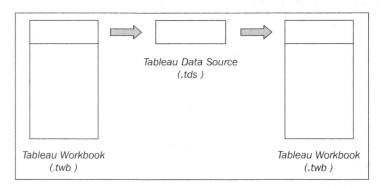

Tableau Data Source
(.tds)

Tableau Workbook
(.twb)

Tableau Workbook
(.twb)

When you save in the default `\My Tableau Repository\Data Source` directory, the connections will appear as shortcut links on the home and **Connect to Data** screens. This makes it easy to create a template data source connection that you can reuse in many workbooks. Additionally, you can share these files with others to give them a customized data source. The `.tds` file is stored as a separate copy of the connection definition. Any changes to a connection in a workbook will not affect the `.tds` file and any changes to the `.tds` file will not affect workbooks that used the `.tds` file to generate a connection.

Another powerful possibility exists for sharing a customized data connection. You can publish a data source connection and publish it to Tableau Server. To accomplish this, right-click on the data source in the **Data** window or select it from the **Data** menu and then select **Publish to Server**. Tableau Server allows you to define user and group permissions to use and edit published data sources.

When you publish a live connection to Tableau Server, any queries to the source are passed through to the underlying data source. When you publish an extracted connection, the extract is hosted on Tableau Server and no connection is needed to the original data source, except to refresh the extract.

An example of customizing and sharing a connection

Now that you've read through the details of editing metadata and sharing a connection, follow these steps to see a simple example:

1. Create a connection to the coffee chain query, as described in *Chapter 1, Creating Your First Visualizations and Dashboard*.

   Drag and drop the **Market** field on **Columns**, **Product** on **Rows**, and **Profit** on **Color** and again on **Label**.

2. Change the mark type to **Square**. The result is a highlighted table.

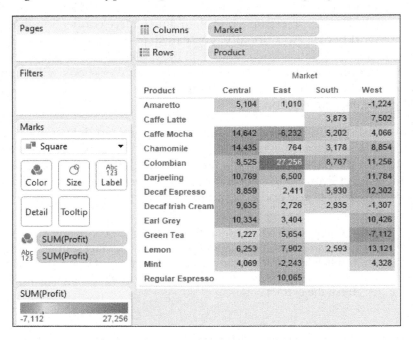

3. Change the default color for **Profit** by right-clicking on the field in the **Data** window and selecting **Default Properties | Color...**. Select the **Orange-Blue Diverging** color palette.

[Orange and blue are considered colorblind-safe. Red and green are often indistinguishable to colorblind individuals.   ]

4. Change the default number format for **Profit** by right-clicking on the field and selecting **Default Properties | Number Format...**. Select **Currency (Custom)**.

5. Change the default sort order of **Market** by right-clicking on the field and selecting **Default Properties | Sort...** and set a **Manual** sort of **East, West, Central**, and **South**.

6. Change the default aggregation of **Profit** by right-clicking on the field and selecting **Default Properties | Aggregation....** Set the default to **Average**. Note that unlike the previous change, this change was not immediately reflected in the view. That's because the active **Profit** fields on **Color** and **Label** were already using an aggregation of **SUM**.

7. Drag and drop the **Profit** field from the **Data** window directly on top of the active **Profit** fields in the view to replace them. Notice that when you drop **Profit** into the view now, the default aggregation is average.

8. Save the data source connection as a `.tds` file by right-clicking on the connection in the **Data** window and selecting **Add to Saved Data Sources....** Save the file as `Custom Coffee Chain Connection.tds` in the `\My Tableau Repository\Data Source` directory. This connection will now show as a shortcut link in Tableau Desktop on the home and **Connection to Data** screens:

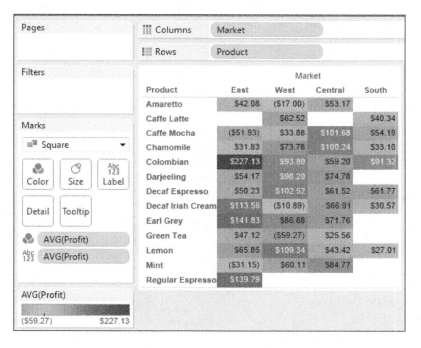

Joins and blends

Joining tables and blending data sources are two different ways to link related data together in Tableau. **Joins** are performed to link tables of data together within a single data source. **Blends** are performed to link together multiple data sources.

Joining tables

Most databases have multiple tables of data that are related in some way. Imagine you have been asked to analyze data in a simple database at a hospital with four main tables, related like this:

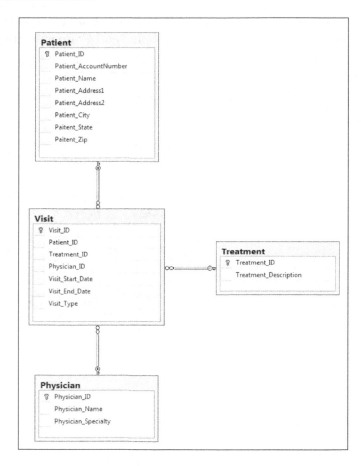

The primary table is the **Visit** table that has a record for every visit of a patient to the hospital and includes details such as the start and end dates and the type of visit (for example, inpatient, outpatient, ER). It also contains key fields that link a visit to a patient, a treatment, and a physician. You have confirmed that every visit has a patient and a physician, but some visits do not necessarily result in a treatment.

When you connect to the database in Tableau, you'll see the four tables listed on the left. Always start by adding the primary table, which in this case is the **Visit** table.

If referential integrity has been defined in the database, Tableau will automatically create the joins as you add additional tables. Otherwise, it will attempt to match field names. In any case, you may adjust the joins as needed. Consider the following joins of the tables in the hospital database:

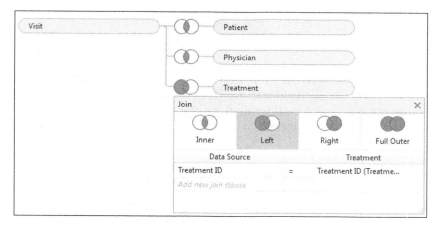

Notice how the small diagram between each table shows what kind of join it is. Clicking on the diagram will allow you to select a different type of join and define which fields are part of the join.

You may specify the following types of joins:

- **Inner**: Only records that match the join condition from both the table on the left and the table on the right will be kept. In this example, only the three matching rows are kept in the results:

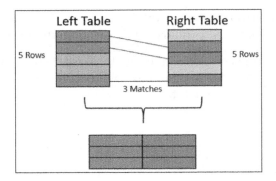

- **Left**: All records from the table on the left will be kept. Matching records from the table on the right will result in values, while unmatched records will contain NULL values for all fields from the table on the right. In this example, the five rows from the left table are kept with NULL results for right values that were not matched:

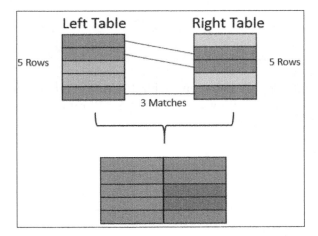

- **Right**: All records from the table on the right will be kept. Matching records from the table on the left will result in values, while unmatched records will contain NULL values for all fields from the table on the left. Not every data source supports a right join. If it is not supported, the option will be disabled. In this example, the five rows from the right table are kept with NULL results for left values that were not matched:

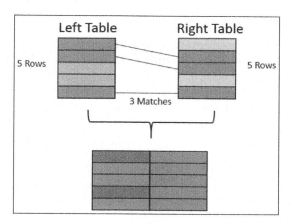

- **Full outer**: All records from tables on both sides will be kept. Matching records will have values from the left and right. Unmatched records will have NULL values where either the left or the right matching record was not found. Not every data source supports a full outer join. If it is not supported, the option will be disabled. In this example, all rows are kept from both sides with NULL values where matches were not found:

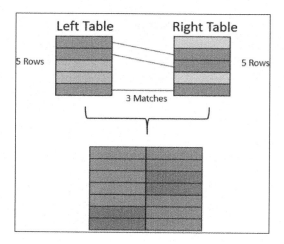

In the hospital example, an inner join was defined for a patient and physician, because every visit (**Left Table**) should find a match with a patient (**Right Table**) and physician (**Right Table**). However, every visit might not have a treatment if, for example, the patient was merely placed in observation. So, a left join was used to ensure that every visit was kept in the results.

Blending data sources

Data blending is a powerful and innovative feature in Tableau. It allows you to use data from multiple data sources in the same view. Often these sources may be of different types. For example, you can blend data from Oracle with data from Excel. You can blend Google Analytics data with that of Access. Data blending also allows you to compare data at different levels of detail. Some advanced uses of data blending will be covered in *Chapter 8, Adding Value to Analysis – Trends, Distributions, and Forecasting*. For now, let's consider the basics and a simple example.

Data blending is done at an aggregate level and involves different queries sent to each data source; unlike joining, which is done at a row level and involves a single query to a single data source. A simple data blending process involves several steps, as shown in the following diagram:

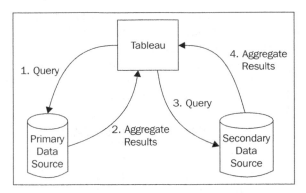

We can see the following from the preceding diagram:

1. Tableau issues a query to the primary data source.
2. The underlying data engine returns aggregate results.
3. Tableau issues another query to the secondary data source. This query is filtered based on the set of values returned from the primary data source for dimensions that link the two data sources.
4. The underlying data engine returns aggregate results from the secondary data source.

Tableau blends the results of the two queries together in the cache. It is important to note how data blending is different from joining. Joins are accomplished in a single query and results are matched row by row. Data blending occurs by issuing two separate queries and then blending together the aggregate results.

There can only be one primary source, but there can be as many secondary sources as you desire. Steps 3 and 4 will be repeated for each secondary source. When all aggregate results have been returned, Tableau will match the aggregate rows based on linking fields.

When you have more than one data source in a Tableau workbook, whichever source you use first in a view becomes the primary source for that view. Blending is view-specific. You can have one data source as the primary in one view and the same data source as the secondary in another. Any data source can be used in a blend, but OLAP cubes, such as SSAS, must be used as the primary source.

Linking fields are dimensions that are used to match data blended between the primary and secondary data sources. Linking fields define the level of detail for the secondary source. Linking fields are automatically assigned if fields match by name and type between data sources. Otherwise, you can manually assign relationships between fields by selecting, from the menu, **Data | Edit Relationships**.

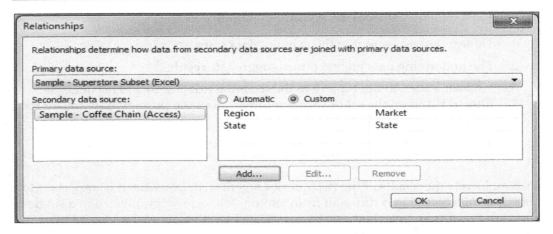

The **Relationships** window will display the relationships recognized between different data sources. You can switch from **Automatic** to **Custom** to define your own linking fields.

Linking fields can be activated or deactivated for blending in a view. Linking fields used in the view will usually be active by default, while other fields will not. You can, however, change whether a linking field is active or not by clicking on the link icon next to a linking field in the **Data** window.

A blending example

The following view shows a simple example of data blending in action:

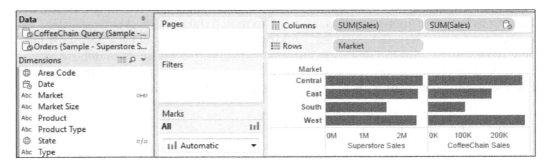

There are two data source connections defined in this workbook, one for the Superstore sample data and the other for the Coffee Chain sample data. The **Superstore** data source is the primary data source in this view (indicated by the blue check mark) and **CoffeeChain** is the secondary source (indicated by the orange check mark). Active fields in the view that are from the secondary data source are also indicated with an orange check mark icon.

The **Sales** measure has been used from both the primary and secondary sources. In both cases, the value is aggregated. The **Market** dimension is an active linking field, indicated by the complete orange link icon next to the field in the **Data** window. The **State** dimension is another linking field, but it is not active in this view.

What this means is that **Sales** is aggregated at the level of **Market** in both cases. In the case of the primary data source, **Market** is a dimension in the view and thus slices the **Sales** measure. In the case of the secondary data source, **Market** defines the level of detail as a linking field.

Data blending will be done based on an exact match of the dimension values. Be careful as this can lead to some matches being missed. For example, if **East** is a value for **Market** in one data source but **east** or **E** in another, the blend will not match the two. You can edit the alias of one of the values so that the two match.

Filtering data

Often, you will want to filter data in Tableau in order to perform analysis on a subset of data, narrow your focus, or drill into detail. Tableau offers multiple ways to filter data.

If you want to limit the scope of your analysis to a subset of data, you can filter the data at the source:

- **Data source filters** are applied before all other filters and are useful when you want to limit your analysis to a subset of data. These filters are applied before any other filters.

- **Extract filters** limit the data that is stored in a TDE file (`.tde`). Data source filters are often converted into extract filters if they are present when you extract the data.

- **Custom SQL filters** can be accomplished using a live connection with custom SQL that has a Tableau parameter in the `WHERE` clause. We'll examine parameters in *Chapter 4, Using Row-level and Aggregate Calculations*.

Additionally, you can apply filters to one or more views using one of the following techniques:

- Drag and drop fields from the **Data** window to the **Filters** shelf.

- Select one or more marks or headers in a view and then select **Keep Only** or **Exclude**.

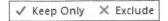

- Right-click on any field in the **Data** window or in the view and select **Show Quick Filter**. Quick filters are filters that are shown as controls (for example, drop-down lists, checkboxes, and so on) to allow the end user of the view or dashboard the ability to change the filter.

- Use an action filter. We'll look more at action filters and quick filters in the context of dashboards.

Each of these options adds one or more fields to the **Filters** shelf of a view. When you drop a field on the **Filters** shelf, you will be prompted with options to define the filter. The filter options will differ most noticeably based on whether the field is discrete or continuous. Whether a field is filtered as a dimension or measure will greatly impact how the filter is applied and the results.

Filtering discrete (blue) fields

When you filter using a discrete field, you will be given options to select individual values to keep or exclude. For example, when you drop the discrete dimension **Department** onto the **Filters** shelf, Tableau will give you the following options:

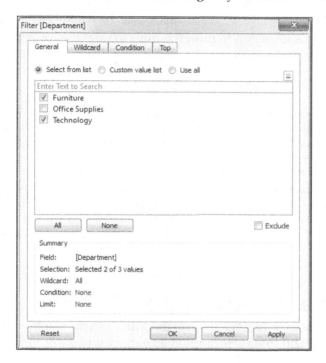

The filter options include the **General**, **Wildcard**, **Condition**, and **Top** tabs. Your filter can include options from each tab. The **Summary** section on the **General** tab will show all options selected. The following is the list of options provided by these tabs:

- The **General** tab allows you to select items from a list (you can use the custom list to add items manually if the dimension contains a large number of values that take a long time to load). You may use the **Exclude** option to exclude the selected items.

- The **Wildcard** tab allows you to match string values that contain, start with, end with, or exactly match a given value.

- The **Condition** tab allows you to specify conditions based on aggregations of other fields that meet conditions (for example, keep any **Departments** where the sum of **Sales** was greater than $1,000,000). Additionally, you can write a custom calculation to form complex conditions. We'll cover calculations more in *Chapter 4, Using Row-level and Aggregate Calculations* and *Chapter 5, Table Calculations*.

- The **Top** tab allows you to limit the filter to only the top or bottom items. For example, you might decide to keep only the top-five items by the sum of **Sales**.

Discrete measures (except for calculated fields using table calculations) cannot be added to the **Filters** shelf. If the field is a date or number, you can convert it to a continuous field before filtering. Other data types will require the creation of a calculated field to convert values you wish to filter into continuous numeric values.

For example, you might have a measure that returns a Boolean (true/false) based on an aggregation and want to filter to keep only the true values. To filter, write a calculated field with code such as IF [Measure] THEN 1 ELSE 0 END and then filter that keeping only values greater than or equal to 1.

Filtering continuous (green) fields

If you drop a continuous dimension onto the **Filters** shelf, you'll get a different set of options. Often, you will first be prompted as to how you want to filter the field:

The options here are divided into two major categories:

- **All values**: The filter will be based on each individual value of the field. For example, an **All Values** filter keeping only **Sales** above $100 will evaluate each record of underlying data and keep only individual sales above $100.

- **Aggregation**: The filter will be based on the aggregation specified (for example, **Sum**, **Average**, **Minimum**, **Maximum**, **Standard deviation**, **Variance**, and so on) and the aggregation will be performed at the **Level of Detail** of the view. For example, a filter keeping only the sum of **Sales** above $100,000 on a view at the level of **Category** will keep only categories that had at least $100,000 in total sales.

Once you've made a selection (or if the selection wasn't applicable for the field selected), you will be given another interface to set the actual filter:

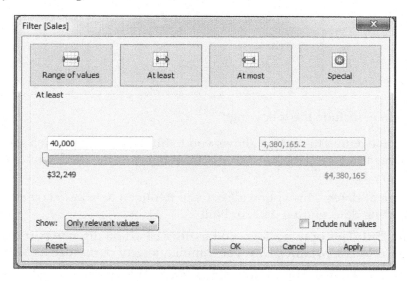

Here, you'll see options to filter continuous values based on a range with a start, end, or both. The **Special** tab gives options to show all values, NULL values, or non-NULL values.

Filtering dates

We'll take a look at the special way Tableau handles dates later. For now, consider the options available when you drop a **Date** field onto the **Filters** shelf:

The options here include the following:

- **Relative date**: This option allows you to filter a date based on a specific date (for example, keep the last 3 weeks from today or the last 6 months from January 1)

- **Range of dates**: This option allows you to filter a date based on a range with a starting date, ending date, or both

- **Date Part**: This option allows you to filter based on discrete parts of dates such as **Years**, **Months**, **Days**, or combinations of parts such as **Month / Year**

- **Individual dates**: This option allows you to filter based on each individual value of the **Date** field in the data

- **Count** or **Count (Distinct)**: This option allows you to filter based on the count or distinct count of date values in the data

Other filtering options

You will also want to be aware of the following options when it comes to filtering:

- You may display a quick filter control for nearly any field by right-clicking on it and selecting **Show Quick Filter**. The type of control depends on the type of field, whether it is discrete or continuous, and may be customized by using the little drop-down arrow in the upper-right area of the quick filter control.

- Filters may be added to the context. Any filters added to the context are evaluated first and a subset of data is stored in a temporary location. Other filters and calculations (such as computed sets) are based on the subset of data. This can be useful if, for example, you want to filter to the top-5 customers but want to be able to first filter to a specific region. Making region a context filter ensures that the top-5 filter is calculated in the context of the region filter.

> Context filters may cause an initial performance hit for a session because Tableau must store the subset of data defined by the context in a temporary location. However, as long as the context doesn't change, all subsequent queries will be made to the context, which may improve subsequent performance.

- Quick filters may be set to show all values in the database, all values in the context, or only values that are relevant based on other filters. These options are available via the drop-down menu on the quick filter control.

- By default, any field placed on the **Filters** shelf defines a filter that is specific to the current view. However, you may right-click on the field and specify the scope. A filter may be applied to all worksheets using a specific data source, a selection of workbooks using the same data source, or a single sheet.

- When using Tableau Server, you may define user filters that allow you to provide row-level security by filtering based on user credentials.

Summary

This chapter covered foundational concepts of how Tableau works with data. Although you will not usually be concerned with what queries Tableau generates to query underlying data engines, having a solid foundational understanding of Tableau's paradigm will greatly aid you as you analyze data. Having a basic understanding of connecting to various data sources, working with extracts, customizing metadata, and the difference between joins and blends will be key as you begin deeper analysis and more advanced visualizations such as those covered in the next chapter.

3
Moving from Foundational to Advanced Visualizations

You are now ready to set out on a journey of building advanced visualizations! "Advanced" does not necessarily mean difficult. Tableau makes many of these visualizations easy to create. Advanced also does not necessarily mean complex. The goal is to communicate the data, not obscure it in needless complexity.

Instead, these visualizations are advanced in the sense that you will need to understand when they should be used, why they are useful, and how to leverage the capabilities of Tableau to create them. Additionally, many of the examples introduce some advanced techniques, such as calculations, to extend the usefulness of foundational visualizations. Many of these techniques will be developed more fully in future chapters, so don't worry about trying to absorb every detail.

Most of the examples in this chapter are designed so that you can follow along. However, don't simply memorize a set of instructions. Instead, take time to understand how the combinations of different field types you place on different shelves change the way headers, axes, and marks are rendered. Experiment and even deviate from the instructions from time to time just to see what else is possible. You can always use Tableau's back button to return to following the example!

Visualizations in this chapter will fall under these major categories:

- Comparison
- Dates and times
- Parts of the whole
- Distributions
- Multiple axes

You may notice the lack of a spatial location or geographic category in the preceding list. Mapping was introduced in *Chapter 1, Creating Your First Visualizations and Dashboard*, and we'll get to some advanced geographic capabilities in *Chapter 10, Advanced Techniques, Tips, and Tricks*.

All the complete examples are included in the workbook of this chapter. You may also use the data sources in this workbook to work through the examples on your own.

Comparing values across different dimensions

Often, you will want to compare the differences of measured values across different dimensions. You might find yourself asking questions like these:

- How much profit did I generate in each department?
- How many views did each of my websites get?
- How many cases did each doctor in the hospital treat last year?

In each case, you are looking to make a comparison (among departments, websites, or doctors) in terms of some quantitative measurement (profit, number of views, and the count of cases).

Bar charts

Here is a simple bar chart, created using the `Superstore Sales` data source, similar to the one we built in *Chapter 1, Creating Your First Visualizations and Dashboard*:

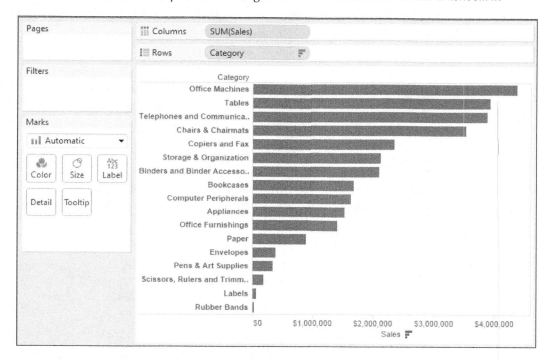

The sum of sales is easily compared for each category of item sold in the chain of stores. **Category** is used as a discrete dimension in the view, which defines row headers (because it is discrete) and slices the sum of sales for each category (because it is a dimension). **Sales** defines an axis (because it is continuous) and is summed for each category (because it is a measure).

Note that the bar chart is sorted with the category that has the highest sum of sales at the top and the lowest at the bottom. Sorting a bar chart adds a lot of value to the analysis because it makes it easy to determine rank. For example, it is easy to see that **Bookcases** has more total sales than **Computer Peripherals** even though the bar lengths are close. Were the chart not sorted, it may not have been as obvious.

You can sort a view in multiple ways:

- Click on one of the sort icons on the toolbar. This results in automatic sorting of the dimension based on the measure that defined the axis. Changes in data or filtering that result in a new order will be reflected in the view.

- Click on the sort icon on the axis. This will also result in automatic sorting.

- Use the dropdown on the active dimension field and select **Sort....** You can also select **Clear Sort** to remove any sorting.

- Drag and drop row headers to manually rearrange them. This results in manual sorting that does not get updated.

Any of these sorting methods are specific to the view and will override any default sort you defined in the metadata.

Bar chart variations

A basic bar chart can be extended in many ways to accomplish various objectives. Consider the following variations:

- Showing progress toward a goal
- Highlighting a single category

Bullet charts – showing progress toward a goal

Let's say you are a global manager and you've set the following profit targets for your regional managers:

Region	Profit target
Central	600,000
East	350,000
South	100,000
West	300,000

You maintain these goals in a spreadsheet and might like to see a visualization that shows you how actual profit compares with your goals. Continue using the workbook with the connection to the Superstore Sales data and follow these steps to see a couple of options:

1. You will find a **Profit Targets** connection in the workbook of this chapter containing this data. The data is also located in a text file in the \Resources\ Data directory of the resources available with this book.

2. Using the Superstore Sales connection, create a basic bar chart showing the sum of profit per region.

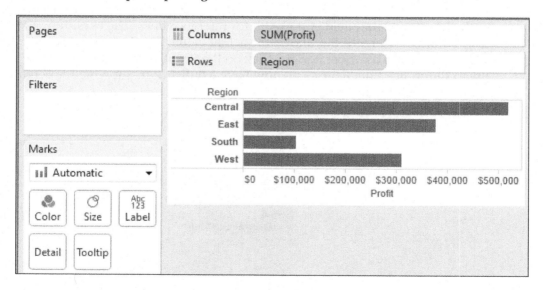

3. Select the **Profit Targets** connection and click to highlight the **Profit Target** field in the **Data** window.

    Open **Show Me** and select the bullet graph. At this point, Tableau has created a bullet graph using the fields in the view and the **Profit Target** field you had selected. You'll observe that the **Reference** field has been used in the data blend to link the two data sources and it is already enabled because the **Region** field was used in the view.

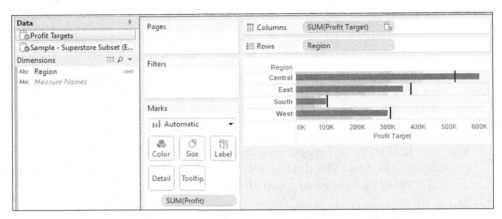

4. Unfortunately, **Show Me** placed **Profit Target** on **Columns** and placed **Profit** in the **Level of Detail** field and used it for reference lines. The bars show the target and the reference line shows the actual value. This is the reverse of what you want. To correct it, right-click on the **Profit Target** axis and select **Swap Reference Line** fields. You now have a bullet chart showing the actual profit compared to the target.

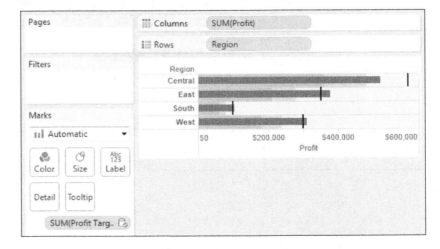

5. Rename this worksheet `Bullet Chart`.

Now, you can clearly see that your **Central** region manager is falling short of the goal you set. Bullet graphs make use of reference lines, which will be more thoroughly covered in later chapters. For now, right-click on the axis and select **Edit Reference Line, Band or Box** to explore this feature.

The bar-in-bar chart

Another way to show the progress toward a goal is to use bar-in-bar charts, such as this:

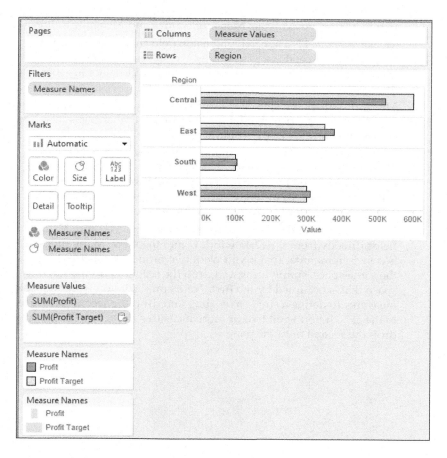

To create this view, continue in the same workbook and follow these steps:

1. Duplicate the `Bullet Chart` sheet you created (right-click on the sheet tab and select **Duplicate Sheet**).

2. Drag the **Profit Target** field from the **Marks** card and drop it directly onto the axis. As you are dropping one measure (**Profit Target**) onto the same space (in this case, an axis) that was being used by another measure (**Profit**), Tableau substituted the special fields' **Measure Names** and **Measure Values**.

Any time you want two or more measures to share the same space within a view, you can use **Measure Names** and **Measure Values**.

Measure Names is a special dimension field that Tableau adds to every data source; it is a placeholder for the names of measures. You can place it in the view anywhere you would place another dimension.

Measure Values is a special measure field that Tableau adds to every data source; it is a placeholder for the values of other measures. You can use it in any way you would use any other measure.

When these special fields are in use, you will see a new **Measure Values** shelf in the workspace. This shelf contains all the measures that are referenced by **Measure Names** and **Measure Values**. You can add and remove measures to and from this shelf as well as rearrange the order of any measures on the shelf.

You can drag and drop the **Measure Names** and **Measure Values** fields directly from the **Data** window into the view. Many times, it is easy to remember that if you want two or more measures to share the same space, simply drag and drop the second onto the same space that is occupied by the first. For example, if you want multiple measures to define a single axis, drag and drop the second measure to the axis. If you want two or more measures to occupy the pane, drop the second onto the pane.

3. Move the **Measure Names** field to the **Color** shelf and edit the colors in the legend (double-click on the legend or use the drop-down arrow on the legend) and set **Profit** to **Orange** and **Profit Target** to **Gray**). You now have a stacked bar chart with a different color for each measure name being used.

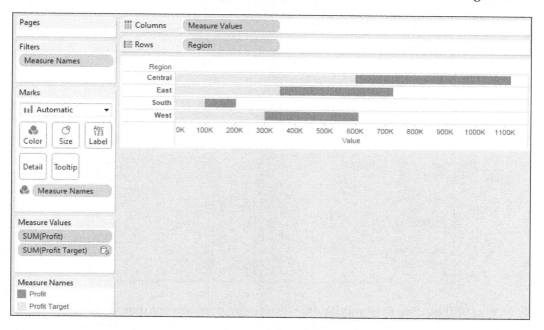

4. Copy the **Measure Names** field from **Color** to the **Size** shelf (hold *Ctrl* while you drag a field in the view and drop it on another shelf in the view). This creates different sizes for each bar segment.

5. Tableau's default is to stack bar marks. In this case, you do not want the bars to be stacked. Instead, you want them to overlap. To change the default behavior, from the menu, navigate to **Analysis | Stack Marks | Off**.

6. Rename the worksheet to Bar-in-Bar.

Highlighting a single category

Let's say one of your primary responsibilities at the superstore is to monitor the sales of tables. You don't necessarily care about the details of other categories, but you do want to keep track of how tables compare with other categories. You might design something like this:

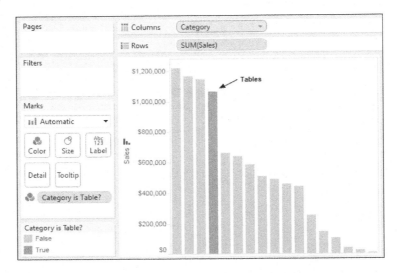

Now, you will be able to immediately see where tables are compared to other categories as sales figures change day to day. To create this view, follow these steps:

1. Place **Category** on **Columns** and **SUM(Sales)** on **Rows**. Sort the bar chart in descending order.

2. Create a calculated field from the menu by navigating to **Analysis | Create Calculated Field...**. Calculations will be covered in much more detail in the next chapter. For now, name the calculation `Category is Table?` and enter the `[Category] = "Tables"` code. This calculation tests the equality and returns `true` if the value of the **Category** field is `"Tables"` and `false` otherwise.

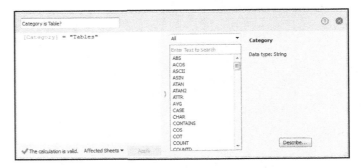

3. When you click on **OK** in the **Calculated Field** dialog, a new field named `Category is Table?` will appear as a new dimension. Drag and drop this field onto the **Color** shelf.

4. Edit the color palette by either double-clicking on the color legend or locating the drop-down caret in the upper-right section of the legend and selecting **Edit** from the drop-down menu. Make the **False** value a light gray.

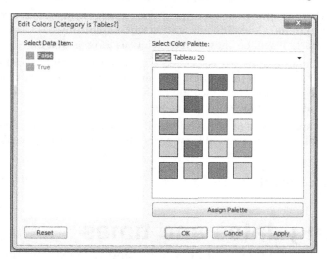

[

You can fine-tune any color and select colors that are not included in a standard palette by double-clicking on the desired value under **Select Data Item** in the **Edit Colors** dialog.
]

5. Hide the column headers for the **Category** field using the drop-down active field menu and unchecking **Show Headers**. As all you care about is how tables relate to other categories and don't specifically care what the other categories are, this will clean up the chart and allow you to focus on what's important.

6. You can add an annotation by right-clicking on the bar and selecting **Annotate | Mark...**.

> Annotations can be used to display values of data and freeform text to draw attention or give explanation. There are three kinds of annotations in Tableau: **Mark**, **Point**, and **Area**.
>
> **Mark** annotations are associated with a specific mark (such as a bar or a shape) in the view. The annotation can display any data associated with the mark. It will be shown in the view as long as that mark is visible.
>
> **Point** annotations are associated with a specific point, as defined by one or more axes in the view. The annotation can display values that define the X and/or Y location of the point. This will be shown in the view as long as the point is visible.
>
> **Area** annotations are associated with an area in the view. They are typically shown when at least part of the defined area is visible.

7. Rename the current sheet `Highlight Single Value`.

Visualizing dates and times

Often in your analysis, you will want to understand when something happened. You'll ask questions like these:

- When did we gain the most new customers?
- What time of the day has the highest call volume?
- What kinds of seasonal trends do we see in sales and profit?

Fortunately, Tableau makes this kind of visual discovery and analysis easy.

The built-in date hierarchy

When you are connected to a flat file, relational, or extracted data source, Tableau provides a robust built-in date hierarchy for any date field.

> Cubes/OLAP connections do not allow Tableau hierarchies. You will want to ensure that all date hierarchies and date values you need are defined in the cube.

To see this in action, continue with the visualization workbook and create a view similar to the one shown in the following screenshot by dragging and dropping **Sales** to **Rows** and **Order Date** to **Columns**:

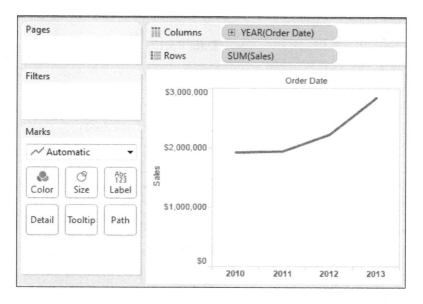

Note that even though the **Order Date** field is a date, Tableau defaulted to showing sales by year. Additionally, the field on **Columns** has a + icon, indicating that the field is part of a hierarchy that can be expanded. When you click on the + icon, additional levels of the hierarchy are added to the view. Starting with **Year**, this includes **Year**, **Quarter**, **Month**, and **Day**. When the field is a date and time, you can further drill down into **Hour**, **Minute**, and **Second**. Any of the parts of the hierarchy can be moved within the view or removed from the view completely.

You may specify how a date field should be used in the view by right-clicking on the date field or using the drop-down menu and selecting various date options.

 As a shortcut, you can right-click and drag drop a date field into the view to get a menu of options for how the date field should be used prior to the view being drawn.

The options for a date field look like this:

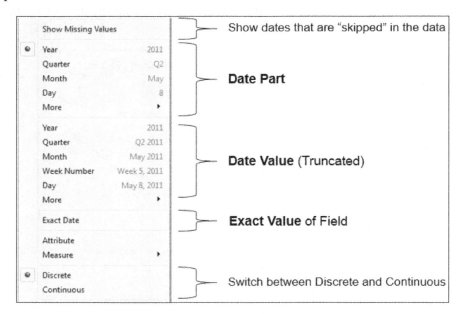

The three major ways a date field can be used are:

- **Date Part**: The field will represent a specific part of the date, such as **Quarter** or **Month**. The part of the date is used by itself and without reference to any other part of the date. This means that a date of November 8, 1980, when used as a month date part, is simply November:

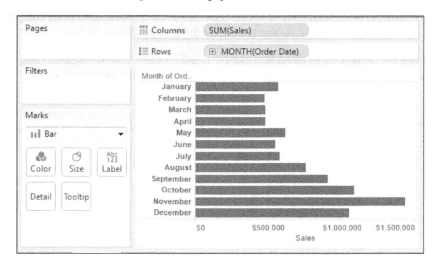

In this view, the bar for **November** represents the sum of sales for all November months regardless of the year or day.

- **Date Value**: The field will represent a date value but rolled up or truncated to the level you select. For example, if you select a date value of **Month**, then November 8, 1980 gets truncated to the month and year and becomes November 1980.

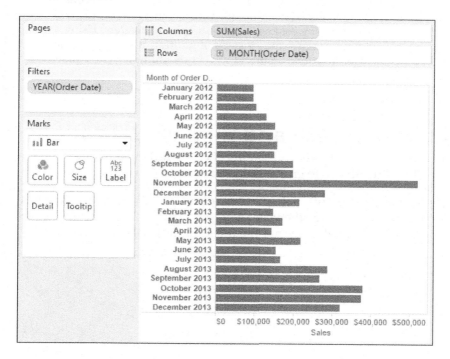

This view, for example, includes a bar for the sum of sales for **November 2012** and another bar for **November 2013**. All individual dates within the month have been rolled up, so sales for November 1, 2013 and November 11, 2013 are all summed under **November 2013**.

- **Exact Date**: The field represents the exact date value (including the time, if applicable) in the data. This means that November 8, 1980 2:01 a.m. is treated as distinct from November 8, 1980 3:08 p.m.

It is important to note that nearly any of these options can be used as discrete or continuous fields. Date parts are discrete by default. Date values and exact dates are continuous by default. However, you can switch them between discrete and continuous as needed to allow flexibility in the visualization.

For example, you must have an axis (and thus, a continuous field) to create a reference line. Also, Tableau will only connect lines at the lowest level of row or column headers. Using a continuous date value instead of multiple discrete date parts will allow you to connect lines across multiple years, quarters, and months.

Variations in date and time visualizations

The ability to use various parts and values of dates and even mix-and-match them gives you a lot of flexibility to create unique and useful visualizations.

For example, using the month date part for columns and the year date part for color gives a time series that makes the year-over-year analysis quite easy. The year date part has been copied to **Label** so that the lines could be labeled.

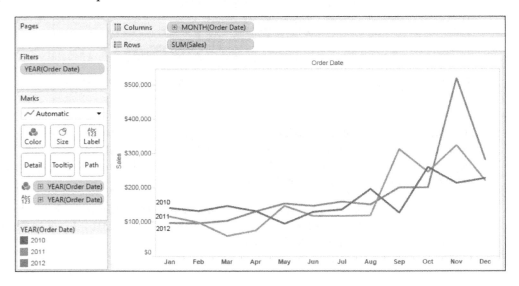

Clicking on any of the shelves on the **Marks** card will give you a menu of options. Here, **Label** has been clicked and the label was adjusted so that it is displayed only at the start of each line.

Here is another example of using date parts on different shelves to achieve useful analysis. This kind of visualization can be quite useful when looking at patterns across different parts of time, such as hours in a day or weeks in a month.

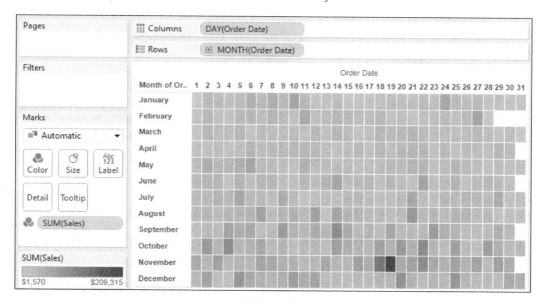

This view shows you the sum of sales for the intersection of each day and each month. **Year** has not been included in the view, so this is an analysis of all years in the data and allows us to see whether there are any seasonal patterns or "hot spots". Observe that placing a continuous field on the **Color** shelf resulted in Tableau completely filling each intersection of row and column with the shade of color that encoded the sum of sales. Clicking on the **Color** shelf gives some fine-tuning options. Here, a white border has been added to help distinguish each cell.

Gantt charts

Gantt charts can be incredibly useful to understand any series of events with duration, especially if these events have some kind of relationship. Visually, they are very useful to determine whether certain events overlap, have dependency, or are longer or shorter than other events. For example, here is a Gantt chart that shows a series of processes, some of which are clearly dependent on others:

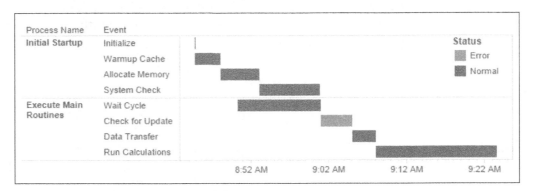

Gantt charts can be created fairly easily in Tableau. Tableau uses the Gantt mark type that places a Gantt bar starting at the value defined by the field defining the axis. The length of the Gantt bar is set by the field on the **Size** card.

Let's say you want to visualize the time it takes from an order being placed to the time the order is shipped. You might follow steps similar to these:

1. Place **Order Date** on **Columns** as **Exact Date** or as a **Day** value. Note that Tableau defaults the marks to Gantt bars.

2. Place **Order ID** on **Rows**. If you receive a warning that there are many members of the dimension, select **Add all Members**. The result is a row for each order. The Gantt bar shows you the date of the order.

> When creating Gantt charts, you will want to include dimensions in the view that give you a meaningful level of detail so that you can see each event of interest. If you are not careful, you could aggregate durations improperly or overlap Gantt marks, resulting in a false representation of data.

3. Filter the view for **December 2013**. Accomplish this by dragging and dropping the **Order Date** field on the **Filters** shelf. Select the **Month/Year** option and then choose the single month and year from the list.

4. The length of the Gantt bar is set by placing a field with a value of duration on the **Size** shelf. There is no such field in this dataset. However, we have the **Ship Date** option and we can create a calculated field for the duration. We'll cover calculations in more detail in the next chapter. For now, select **Analysis** from the menu and click on **Create Calculated Field....** Name the field Time to Ship and enter the following code:

```
DATEDIFF('day', [Order Date], [Ship Date])
```

> When using a date axis, the length of Gantt bars always needs to be in terms of days. If you want to visualize events with durations that are measured in hours or seconds, avoid using the 'day' argument for DATEDIFF because it computes whole days and loses precision in hours and seconds.
>
> Instead, calculate the difference in hours or seconds and then convert it to days. The following code converts the number of seconds between a start and end date and then divides it by 86,400 to convert the result into days (including fractional parts of the day):
>
> ```
> DATEDIFF('second', [Start Date], [End Date]) / 86400
> ```

5. The new calculated field should appear under **Measures** in the **Data** window. Drag and drop the field onto the **Size** shelf. You now have a Gantt chart showing when orders were placed. There is, however, one problem. Some orders include more than one item and you are showing the sum of days to ship. This means that if one item took 5 days to ship and another item in the same order took 7 days, the length of the bar shows 12 days for the order. If both items took 5 days and were shipped at the exact same time, the length of the bar indicates 10 even though the order really only took 5 days.

6. To correct this, decide whether you want to show the minimum number of days or the maximum number of days for each order, right-click on the **Days to Ship** field on the **Marks** card, or use the drop-down menu and navigate to **Measure | Minimum** or **Measure | Maximum**. Alternately, you might decide to add **Item** to the **Detail** card of the **Marks** card.

Your final view should look something like this:

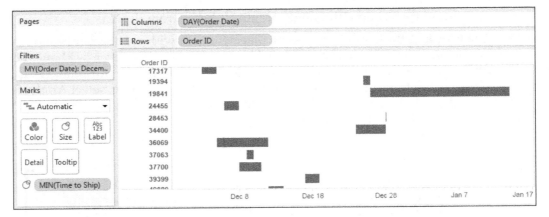

> Often, you'll want to sort a Gantt chart so that the earliest start dates appear first. Do this via the drop-down menu of the dimension on **Rows** and select **Sort**. Sort it in ascending order by the minimum of the date field.

Relating parts of the data to the whole

As you explore and analyze data, you'll often want to understand how various parts add up to a whole. For example, you'll ask questions like these:

- How many patients with different admission statuses (in-patient, out-patient, observation, or ER) make up the entire population of patients in the hospital?
- What is the percentage of total national sales made in each state?
- How much space does each file, subdirectory, and directory take on my hard disk?

These types of questions ask about the relationship between the part (patient type, state, or file/directory) and the whole (the entire patient population, national sales, and hard disk). There are several types of visualizations and variations that can aid you in your analysis.

Stacked bars

We took a look at stacked bars in *Chapter 1, Creating Your First Visualizations and Dashboard*, where we noted one significant drawback: it is difficult to compare values across categories for any but the bottom-most bar (for vertical bars) or the left-most bar (for horizontal bars). The other bar segments have different starting points, so lengths are much more difficult to compare.

In this case, however, we are using stacked bars to visually understand the makeup of the whole. We are less concerned with a visual comparison across categories.

Say a bank manager wants to understand the makeup of her lending portfolios. She might start with a visualization like this:

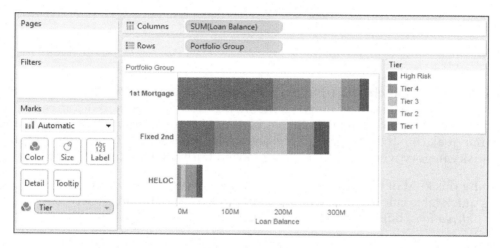

This gives a decent view of the makeup of each portfolio. However, in this case, the bank manager already knows that the bank has more balance in first-mortgage loans than fixed second loans. However, she wants to understand whether the relative makeup of the portfolios is similar; specifically, do the **High Risk** balances constitute a higher percentage of balances in any portfolio?

Consider this alternative:

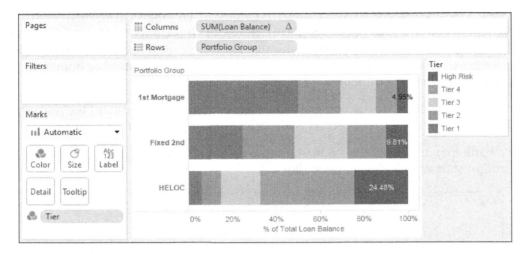

None of the data has changed, but the bars now represent the percent age of the total of each portfolio. You can no longer compare the absolute values, but comparing the relative breakdown of each portfolio has been made much easier. The bank manager may find it alarming that nearly 25 percent of the balance of HELOC loans is in the high-risk category when the bar segment looked fairly small in the first visualization.

Creating this kind of visualization is relatively easy in Tableau. It involves using quick table calculations, which will be covered in depth in *Chapter 5, Table Calculations*, but it only takes a few clicks to implement.

Continuing with the `Advanced Visualizations` workbook, follow these steps:

1. Create a stacked bar chart like what is shown in the following screenshot with **Department** on **Rows**, **Shipping Cost** on **Columns**, and **Ship Mode** on **Color**.

2. Duplicate the **Shipping Cost** field on **Columns** either by holding *Ctrl* while dragging the **Shipping Cost** field from **Columns** to **Columns**, immediately to the right of its current location, or by dragging and dropping it from the **Data** window to **Columns**. At this point, you have two **Shipping Cost** axes that, in effect, duplicate the view.

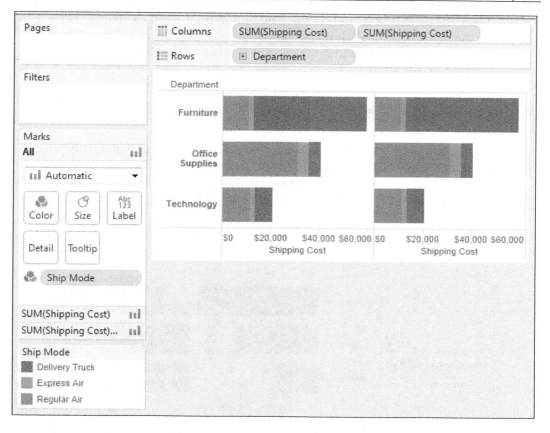

3. Using the drop-down menu of the second **Shipping Cost** field, navigate to **Quick Table Calculation | Percent of Total**. This table calculation runs a secondary calculation on the values returned from the data source to compute a percent of the total. You will need to further specify how that total should be computed.

4. Using the same drop-down menu, navigate to **Compute Using | Table (Across)**. This tells Tableau to compute the table calculation across the table, which in this case means that the values will add up to 100 percent for each department.

5. Turn on labels by clicking on the **Abc** button on the top toolbar. This turns on default labels for each mark. As you've already seen, you can customize labels by dropping a field on the **Label** shelf and fine-tune it further by clicking on the shelf.

6. Right-click on the second axis, which is now labeled % **of Total Shipping Cost** and select **Edit Axis...**. Then, set **Range** as **Fixed** from **0** to **1**. In this case, you know the total will always be **100%**, so fixing the axis from 0 to 1 allows Tableau to draw the bars all the way across.

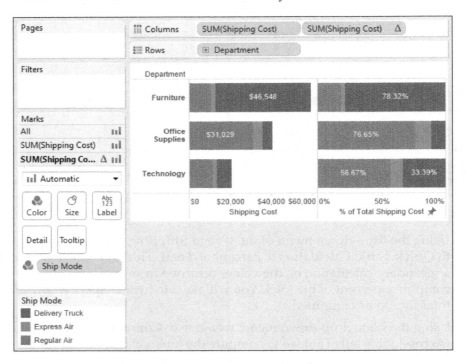

Treemaps

Treemaps use a series of nested rectangles to represent hierarchical relationships of parts to whole. Treemaps are particularly useful when you have hierarchies and dimensions with high cardinality (a high number of distinct values).

Here is an example of a treemap that shows you how the sales of each item add up to give the total sales by category, then department, and finally, the total sales overall. Profit has been encoded by color to add additional analytical value to the visualization. It is now easy to pick out items with negative profit that have relatively high sales when placed in the context of the whole:

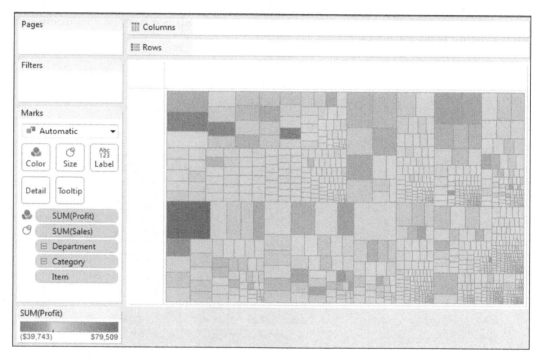

Treemaps, along with packed bubbles, word clouds, and a few other chart types, are called **non-Cartesian** chart types. This means that they are drawn without an *x* or *y* axis and do not even require row or column headers.

To create a treemap, you simply need to place a measure on the **Size** shelf and a dimension on the **Detail** shelf. You can add additional dimensions to the level of detail to increase the detail of the view.

You can quickly change a treemap into a word cloud or a packed bubble chart by changing the mark type from **Automatic** (which is **Square**) to **Circle** (for packed bubbles) or **Text** (for word clouds).

The order of the dimensions on the **Marks** card defines the way the treemap groups the rectangles. Additionally, you can add dimensions to rows or columns to slice the treemap into multiple treemaps. Effectively, the end result is a bar chart of treemaps! The following is an example:

The treemap in the preceding screenshot not only demonstrates the ability to have multiple rows (or columns) of treemaps, but it also demonstrates the technique of placing multiple fields on the **Color** shelf. This can only be done with discrete fields. You can assign two or more colors by holding the *Shift* key while dropping the second field on the color. Alternately, the icon or space to the left of each field on the **Marks** card can be clicked on to change which shelf is used for the field.

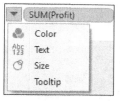

Area charts

You might think of an area chart as a line chart in which one line is drawn with the area under it filled. Subsequent areas are stacked on top.

As an example, consider a visualization of delinquent loan balances being analyzed by the bank manager:

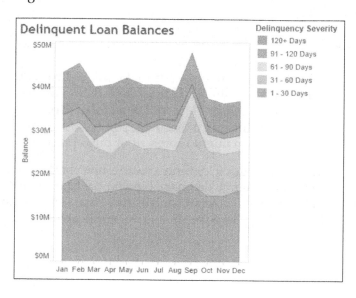

This area chart shows you the delinquent balance over time. Each band represents a different severity of delinquency. In many ways, the view is aesthetically pleasing, but it suffers from some of the same weaknesses as the stacked bar chart. As all but the bottom band have different starting locations month to month, it is difficult to compare the bands between months. For example, it is obvious that there is a spike in delinquent balances in September. But is it in all bands? Or is one of the lower bands pushing the higher bands up? Which band has the most significant spike?

Now, consider this similar view:

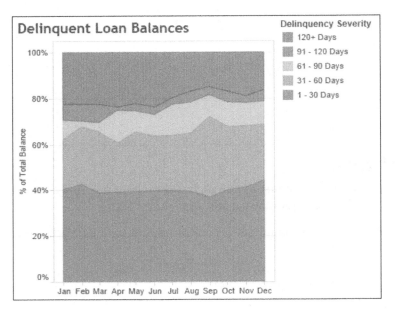

This view uses a quick table calculation similar to the stacked bars. It makes it clearer that the percent age of balance within the 31 to 60 days delinquent range increased in September. However, it is no longer clear that September represents a spike in balances. If you were telling a story with this data, you would want to carefully consider what either visualization might represent or misrepresent.

Creating an area chart is fairly simple. Simply create a line chart or time series as you did previously, and then change the mark type on the **Marks** card to **Area**. Any dimensions on the **Color**, **Label**, or **Detail** shelves will create slices of area that will be stacked on top of each other. The **Size** shelf is not applicable to an area chart.

You can define the order in which the areas are stacked by changing the sort order of the dimensions on the shelves of the **Marks** card. If you have multiple dimensions defining slices of area, you can also rearrange them on the **Marks** card to further adjust the order.

Pie charts

Pie is actually a mark type in Tableau. As we'll see in future chapters, this gives you some additional flexibility with pie charts that is not available for other chart types, such as the ability to place them on maps.

Creating a pie chart is not difficult. Simply change the mark type to **Pie**. This will give you an **Angle** shelf that you can use to encode a measure. Whatever dimension(s) you place on the **Marks** card (typically on the **Color** shelf) will define the slices of the pie. The following is an example of a pie chart in Tableau:

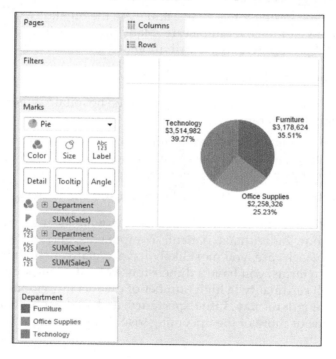

You'll notice that the pie chart here uses **Sales** to define the angle of each slice; the higher the sum of sales, the wider the slice. The **Department** dimension slices the measure and defining slices of the pie. This view also demonstrates the ability to place multiple fields on the **Label** shelf. The second **SUM(Sales)** field is the percent age of the total table calculation you saw previously.

Be careful when using pie charts. Most visualization experts will affirm that it is far more difficult for the human eye to differentiate differences in angles than it is to differentiate differences in length or position. For example, without the labels in the preceding chart, would you really be able to tell whether one slice was really 25 percent instead of 30 percent? A bar chart showing sales for the three departments would be more readable.

Visualizing distributions

Often, simply understanding totals, sums, and even the breakdown of part-to-whole only gives you a piece of the overall picture. Many times, you'll want to understand where individual items fall within a distribution of all similar items.

You might find yourself asking questions like these:

- How long do most of our patients stay in the hospital? Which patients fall outside the normal range?

- What's the average life expectancy for components in a machine and which components fall above or below that average? Are there any components with extremely long or extremely short lives?

- How far above or below "passing" were most students' test scores?

These questions all have similarities. In each case, you are asking for an understanding of how individuals (patients, components, and students) compared with each other. In each case, you most likely have a relatively high number of individuals. In data terms, you have a dimension (**Patient**, **Component**, and **Student**) with high cardinality (a high number of distinct individual values) and some measures (**Length of Stay**, **Life Expectancy**, and **Test Score**) you'd like to compare. Using one or more of the upcoming visualizations might be a good way to do this.

Circle charts

Circle charts are one way to visualize a distribution. Consider the following view, which shows you how each state compares to other states within the same region in terms of total profit:

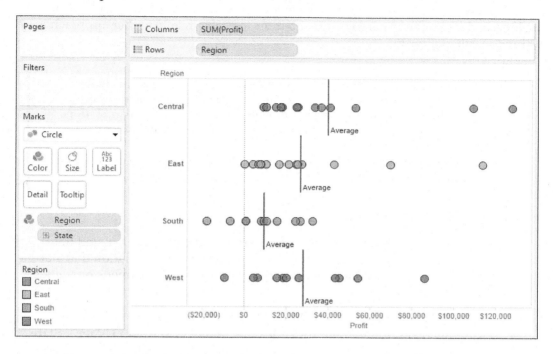

Here, you can easily see that certain states do far better or far worse than others in terms of profit. More than that, you can see whether the state has made or lost money and how much above or below the regional average the state was.

Creating the view is not difficult. After placing the fields on shelves as shown in the preceding screenshot, simply change the mark type from **Automatic (Bar)** to **Circle**. **Region** defines the rows and each circle is drawn at the level of state that is in the level of **Detail** on the **Marks** card. Finally, to add the average lines, simply right-click on the **Profit** axis and select **Add a Reference Line, Band, or Box....** In the resulting options window, add a line, **Per Cell**, for the average of **SUM(Profit)**. You can adjust the label and formatting of the reference line as desired.

Jittering

When using views such as circle plots, you'll often see that marks overlap, which can lead to obscuring of the true story. Do you know for certain, just by looking, that there is only one state in the west region that is unprofitable? Or could there be two circles exactly overlapping? One way to minimize this is to click on the **Color** shelf and add some transparency and a border to each circle. Another approach is a technique called jittering.

Jittering is a common technique in data visualization that involves adding a bit of intentional noise to a visualization to avoid overlap without harming the integrity of what is communicated. Alan Eldridge and Steve Wexler are among those who pioneered techniques for jittering in Tableau.

Other jittering techniques can be found by searching for `jittering` on the Tableau forums or `Tableau jittering` using a search engine.

Here is one approach to add a jitter:

1. Start by duplicating the view created earlier and create a calculated field named `Index` with the `Index()` code. Index is a table calculation that starts with a value of 1 and increments for each intersection of dimension values within a partition of the view.

2. Place the newly created **Index** field on **Rows** to the right of **Region**.

3. Using the drop-down menu for **Index**, navigate to **Compute Using | State**.

4. Using the drop-down menu for **Index**, uncheck **Show Header** to hide the axis. The values do not carry any meaning that needs to be conveyed to your audience.

5. The faint horizontal grid lines do not add anything to the visualization. From the menu, navigate to **Format | Lines** and then set **Grid Lines** to **None**. Alternately, you may choose to keep the vertical lines, so instead, set **Grid Lines** to **None** under the **Rows** tab only. The following will be the result:

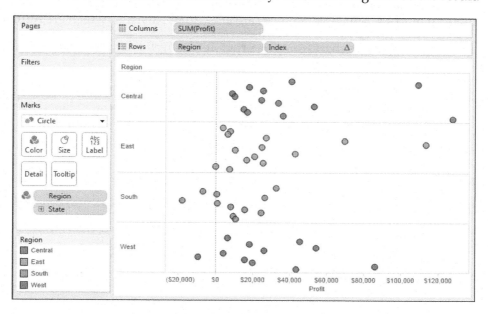

What you've done is index each state within each region. As **Index** is continuous (green), it defines an axis and causes the circles to *spread out* vertically. Now you can more clearly see each individual mark and have higher confidence that overlap is not obscuring the true picture of the data. You can use jittering techniques on many different kinds of visualizations.

Box and whisker plots

Box and whisker plots add additional information and context to distributions. They show the upper and lower quartile and whiskers, which extend to either 1.5 times the upper/lower quartile or to the maximum/minimum values in the data. This allows you to see which data points are close to normal and which are outliers.

The following is the circle chart from the previous example, with the addition of boxes and whiskers:

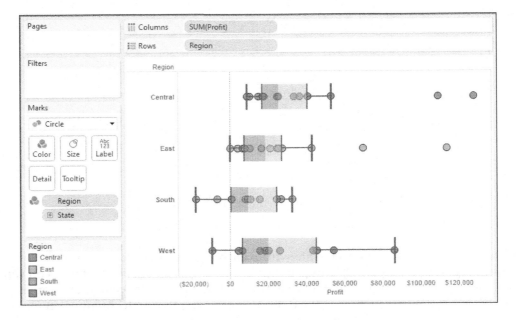

To add box and whisker plots, right-click on an axis and select **Add a Reference Line, Band, or Box...**. Select **Box Plot** and set the desired options and formatting.

Histograms

Another possibility to show a distribution is to use a histogram. A histogram looks similar to a bar chart, but the bars show the count of occurrences of a value. For example, standardized test auditors looking for evidence of grade tampering might construct a histogram of student test scores.

Typically, a distribution might look like this:

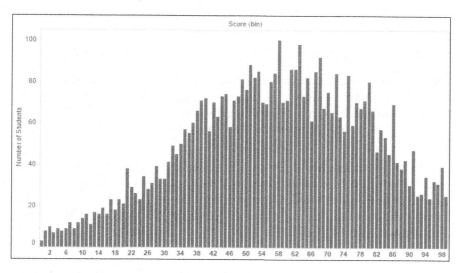

The test scores are shown on the x axis and the height of each bar shows the number of students who had that particular score. A typical distribution should have a fairly recognizable curve with some students doing poorly, some doing extremely well, and most falling somewhere in the middle.

Consider the implications if auditors observed the following visualization:

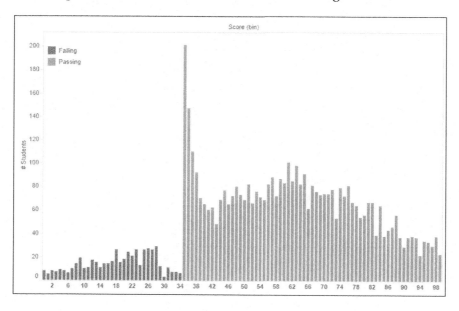

Something is clearly wrong. It appears that graders bumped up students who were just shy of passing to *barely passing*. Histograms are very useful in catching anomalies like this.

You can create a histogram in Tableau by following steps similar to these:

1. Create a **bin** for a numeric field (typically, a measure) by right-clicking on the field in the **Data** window and selecting **Create Bins…**. A bin is a grouping of measure values by range. You can define the size of each bin. The bin will be displayed as a new **Dimension** field in the **Data** window. For this example, create bins of 100 for the **Sales** measure.

2. Based on your data and questions, decide what you want to count. Do you want to count the number of records that occur in each sales bin? Use the **Number of Records** field. Do you want to count the number of distinct customers that fall into each bin? Right-click plus drag the **Customer ID** field onto **Rows** and select **CNTD**.

Histograms can also be created very easily using **Show Me**. Simply select a single measure and then select **Histogram** from **Show Me**. It will create the bin and place the required fields on the view. You can adjust the size of a bin by right-clicking on it in the **Data** window.

Here is an example of a histogram of the number of distinct customers for each sales bin. More customers purchased between $0 and $100 than any other range.

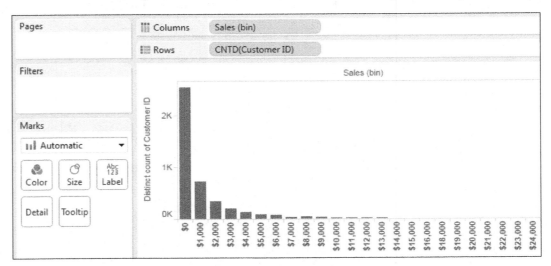

Just like dates, the **bin** field drop-down includes an option for **Show Missing Values**. This can be very useful to avoid distorting the visualization and to identify what values *don't* occur in the data.

Visualizing multiple axes to compare different measures

Often, you'll need to use more than one axis to compare different measures, understand correlation, or analyze the same measure at different levels of detail. In these cases, you'll use visualizations with more than one axis.

Scatterplots

A scatterplot is an essential visualization type to understand the relationship between two measures. Consider a scatterplot when you find yourself asking questions like these:

- Does how much I spend on marketing really make a difference to sales?
- How much does power consumption go up with each degree of heating/cooling?
- Is there any correlation between rental price and the length of contract?

Each of these questions seeks to understand the correlation (if any) between two measures. Scatterplots are great to see these relationships and also to locate outliers.

Consider the following scatterplot, which looks at the relationship between the measures of the sum of **Sales** (on the *x* axis) and the sum of **Profit** (on the *y* axis):

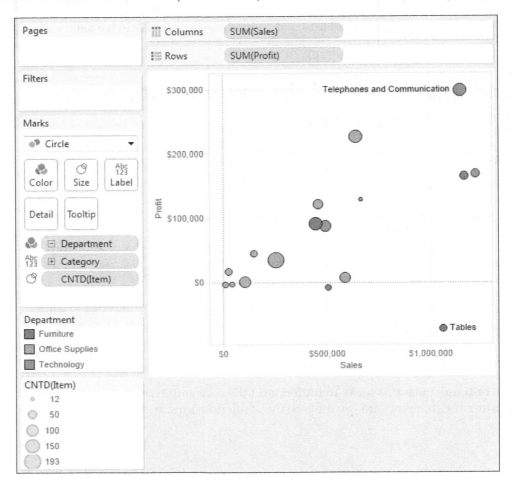

The dimensions of **Department** and **Category** on the **Marks** card define the level of detail. **Color** has been used to make it easy to see which department the category belongs to. Each mark in the view represents the total sales and total profit for a particular category in a particular department. The size of each circle indicates the number of distinct items sold in the category/department. The scatterplot points out an issue with tables. They have high sales but are unprofitable. Telephones, on the other hand, have high sales and high profit.

Dual Axis

One very important features of Tableau is **Dual Axis**. Scatterplots use two axes, and they are *x* and *y*. You've already seen how to use **Measure Names** and **Measure Values** to show more than one measure on a single axis. You saw in the stacked bar example that placing multiple continuous (green) fields next to each other on **Rows** or **Columns** results in multiple side-by-side axes. Dual axis, on the other hand, means that a view is using two axes that are opposite to each other with a common pane.

For example, this view uses a dual axis for **Sales** and **Profit**:

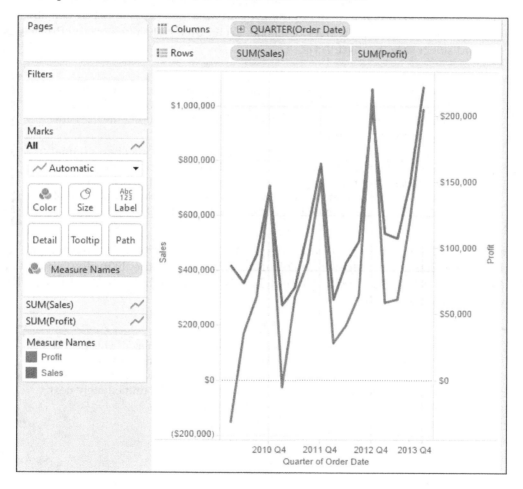

You can observe several key features of the view:

- The **Sales** and **Profit** fields on **Rows** indicate that they have a dual axis by sharing a flattened side.

- The **Marks** card is now an accordion-like control with an **All** section and a section for **Sales** and **Profit**. You can use this to customize marks for all measures or specifically customize marks for either **Sales** or **Profit**.

- **Sales** and **Profit** both define *y* axes that are on opposite sides of the view.

- Note that the peaks of the lines might lead you to believe that **Sales** and **Profit** were roughly equal in value at certain points. This is unlikely to be the case. Indeed, it is not. Instead, the axes are not in sync. Sales of $1,000,000 align with profit of just over $200,000. To fix this, right-click on the **Profit** axis and select **Synchronize Axis**.

You must set the synchronize option using the secondary axis (**Profit** in the example). If the **Synchronize Axis** option is ever disabled on the secondary axis, it is likely that the two fields defining the axes are different numeric types.

For example, one may be an integer, while the other may be a decimal. To enable the synchronize option, you'll need to force a match of the types by either changing the data type of one of the fields in the metadata or by creating a calculated field that specifically casts one of the fields to the matching type (for example, **INT** or **FLOAT**).

Creating a dual axis is relatively easy. Simply drag and drop two continuous (green) fields next to each other on **Rows** or **Columns**, and then use the drop-down menu on the second and select **Dual Axis**. Alternately, you can drop the second field onto the canvas opposite the existing axis.

Dual axes can be used with any field that defines an axis. This includes continuous numeric fields, continuous date fields, and latitude or longitude fields that define a geographic visualization. In the case of latitude or longitude, simply copy the field, place it immediately next to itself on **Rows** or **Columns**, and select **Dual Axis**.

Combination charts

Combination charts extend the use of dual axes to overlay different mark types. This is possible because the **Marks** card will give options to edit all marks or customize marks for each individual axis.

 Multiple mark types are available any time two or more continuous fields are located beside each other on **Rows** or **Columns**. This means that you can create views with multiple mark types even when you are not using a dual axis.

Consider the following visualization:

This chart uses a combination of bars and lines to show the total sales over time (using bars) and the breakdown of sales by department over time (using lines). This kind of visualization can be quite effective at giving additional context to detail.

There are several things to note about this view:

- The field on the **Color** shelf is listed as **Multiple Fields** and is gray on the **Marks** card. This indicates that different fields have been used for **Color** for each axis on the **Marks** card (no field for the first **Sales** axis and **Department** for the second).

- The view demonstrates the ability to mix levels of detail in the same view. The bars are drawn at the highest level, while the lines are drawn at the level of **Department**.

- The view demonstrates the ability to use the same field (**Sales**, in this case) multiple times on the same shelf (**Rows**, in this case).

- The second axis (the **Sales** field on the right-hand side) has the header hidden to remove redundancy from the view.

- The months have been formatted to show abbreviations. This was done via the drop-down menu of the **Order Date** field on **Columns**, selecting **Format**, and selecting the desired format of the field for headers.

Dual axis and combination charts open a wide range of possibilities for mixing mark types and levels of detail. We'll see a few more examples throughout the rest of the book, but you should definitely experiment with this feature and let your imagination run wild with all that can be done.

Summary

We covered quite a bit of ground in this chapter! You should now have a good grasp of when to use certain types of visualizations. The types of questions you ask of the data will often lead you to a certain type of view. You explored how to create these various types and how to extend basic visualizations using a variety of advanced techniques, such as calculated fields, jittering, multiple mark types, and dual axis. Along the way, we also covered some details on how dates work in Tableau using the special **Measure Names / Measure Values** fields.

Hopefully, the examples using calculations have made you eager to learn more about creating calculated fields. The ability to create calculations in Tableau opens up endless possibilities for extending data, calculating results, customizing visualizations, and creating rich user interactivity. We'll dive deep into calculations in the next two chapters to see how they work and what amazing things they can do.

4
Using Row-level and Aggregate Calculations

One of the most incredible things about Tableau is that it is intuitive and transparent to use. We have already seen what amazing discovery, analysis, and data storytelling is possible in Tableau by simply connecting to data and dragging and dropping fields.

Calculations significantly extend the possibilities for analysis, design, and interactivity in Tableau. In this chapter, we'll see how calculations can be used in many ways. We'll see how calculations can be used to fix common problems with data, extend the data by adding new dimensions and measures, and provide additional flexibility in interactivity.

At the same time, while calculations provide incredible power and flexibility, they introduce a level of complexity and sophistication. As you work through this chapter, seek to understand the key concepts behind how calculations work in Tableau. As usual, follow along with the examples, but feel free to explore and experiment. The goal is not to merely have a list of calculations you can copy, but to gain knowledge of how calculations can be used to solve problems and add creative functionality to your visualizations and dashboards.

In this chapter, we'll look at two of the three levels of calculations in Tableau: row level and aggregate calculations. The first half of the chapter focuses on some foundational concepts while the last half gives quite a few practical examples. The topics include:

- Creating calculations
- Overview of the three levels of calculation
- Parameters
- Practical examples
- Performance considerations

We'll examine table calculations in the next chapter.

The examples in this chapter will use the following dataset. It's simple and small so that we can easily see how the calculations are being done. This dataset is included as `Rental Data.xls` in the `Learning Tableau\Data` directory of the book resources and is also included in the workbook of this chapter as a data source named `Rental Data`.

Apartment	Renter	Start	End	Area	Price
1	Steve Agee	03.17.2014	Oct-14	1,000	5,000
2	Jason Bush	01.17.2014	Mar-14	500	2,000
3	Matthew Lederman	04.01.2014	Jul-14	500	2,500
4	Nathan Mackenroth	11.08.2014	Dec-14	1,000	4,000
5	Jim Mihalick	07.17.2014	Nov-14	500	5,000
6	Charlie Williams	05.05.2014	Aug-14	2,000	2,000

The dataset describes six apartments, the renter, the start and end dates of the rental period, the area (in square feet), and the monthly rental price.

Creating and editing calculations

A calculation is often referred to as **Calculated Field** in Tableau. This is because when you create a calculation, it will show up as either a new measure or dimension in the **Data** window. Calculations consist of code that reference other fields, parameters, constants, or sets and use combinations of functions and operations to achieve a result. Sometimes, this result is per row of data and sometimes it is done at an aggregate level. We'll consider the difference shortly.

There are multiple ways to create a calculated field in Tableau:

- Select **Analysis | Create Calculated Field...** from the menu
- Right-click on an empty area in the **Data** window and select **Create Calculated Field**
- Right-click on a field, set, or parameter and select **Create | Calculated Field...**
- In Tableau 9.0 or later, double-click on an empty area on the **Rows, Columns,** or **Measure Values** shelves, or in the empty area on the **Marks** card to create an ad hoc calculation

When you start your calculation from an existing field or parameter, the calculation starts as a reference to that field. The calculated field you create will be part of the data source that is currently selected at the time you create it. You can edit an existing calculated field by right-clicking on it in the **Data** window and selecting **Edit...**.

The interface to create and edit calculations looks like this:

The window has several key features:

- **1**: Once created, the calculated field will show up as a field in the **Data** window (for the data source indicated) with the name you enter.
- **2**: The code editor allows you to type in the code for the calculation. The editor includes autocomplete for recognized fields and functions. Additionally, you may drag fields, sets, and parameters from the **Data** window into the code editor to insert them into your code.

You may select snippets of your code in the window and then drag and drop the selected text into the **Data** window to create additional calculated fields. You may also drag and drop selected code snippets from the code window onto shelves in the view to create ad hoc calculations. This is an effective way to test portions of complex calculations.

- **3**: An indicator at the bottom of the editor will alert you to errors in your code. Additionally, you can view which sheets will be affected by changes to the calculation.

- **4**: The functions list contains all the various functions available for use in your code. Many of these functions will be used in examples or discussed in this chapter. Tableau groups various functions according to their overall use:

  - **Number**: This includes mathematical functions such as rounding, absolute value, trig functions, square roots, and exponents.

  - **String**: This includes functions useful for string manipulation such as getting a substring, finding a match within a string, replacing parts of a string, and converting a string value to uppercase or lowercase.

  - **Date**: This includes functions useful for working with dates such as finding the difference between two dates, adding an interval to a date, getting the current date, and transforming strings with nonstandard formats to dates.

  - **Type conversion**: This includes functions useful for converting one type of field to another, such as converting integers to a string, floating point decimals to integers, or strings to dates.

  - **Logical:** This includes decision-making functions such as `if then else` logic or `case` statements.

  - **Aggregate**: This includes functions used to aggregate such as summing, getting the minimum or maximum values, or calculating standard deviations or variances.

  - **User**: This includes functions used to obtain usernames and check whether the current user is a member of a group. These functions are often used in combination with logical functions to customize the user's experience or to implement user-based security when publishing to Tableau Server or Tableau Online.

  - **Table calculation**: These functions are different from others. They operate on the aggregate data *after* it is returned from the underlying data source and just prior to the rendering of the view. These are some of the most powerful functions in Tableau. We'll devote an entire chapter to cover them.

- **5**: Selecting a function in the list or clicking on a field, parameter, or function in the code will reveal details about the selection on the right. This is helpful when nesting other calculated fields in your code and you want to see the code for that particular calculated field or when you want to understand the syntax for a function.

As you edit, you may click on **Apply** to refresh the view with the changes you've made. **OK** will apply the changes and close the editor. The **X** button in the upper-left area of the editor will close the editor without saving any unapplied changes.

Tableau supports numerous functions and operators. In addition to the functions listed on the calculation screen, Tableau supports the following operators, key words, and syntax conventions:

- **AND**: This is a logical "and" between two Boolean (`true`/`false`) values or statements
- **OR**: This is a logical "or" between two Boolean values or statements
- **NOT**: This is a logical "not" to negate Boolean values or statements
- **= or ==**: These are a logical "equals" to test the equality of two statements or values
- **+**: This represents the addition of numeric or date values, or concatenations of strings
- **-**: This represents the subtraction of numeric or date values
- *****: This represents the multiplication of numeric values
- **/**: This represents the division of numeric values
- **^**: This raises to a power with numeric values
- **()**: Parentheses define the order of operations
- **[]**: Square brackets are used to enclose field names
- **{}**: Curly Braces are used to enclose **Level of Detail** calculations
- **//**: Double slash is used to start a comment

Field names that are a single word may optionally be enclosed in brackets when used in calculations. Field names with spaces, special characters, or from secondary data sources must be enclosed in brackets.

Three levels of calculation

The groupings of functions mentioned earlier are important for understanding what kind of functionality is possible. However, the most fundamental way to understand calculations in Tableau is to think of three different levels of calculation:

- **Row-level calculations**: These calculations are performed for every row of underlying data
- **Aggregate-level calculations**: These calculations are performed at an aggregate level, which is usually defined by the dimensions used in the view
- **Table calculations**: These calculations are performed on the table of aggregate data that has been returned by the data source to Tableau

Row-level and aggregate-level calculations are processed as part of the query executed by the underlying source data engine. Row-level calculations are computed for each row of the source. Aggregate calculations are also computed at the data source, but the calculation is at an aggregate level. Table calculations are not calculated as part of the query to the data source. Instead, when the data source returns aggregate results, those results are stored in the cache. Table calculations are applied in the cache just prior to the view being rendered.

Understanding these three levels will make working with calculations much more pleasant. Let's consider some basic examples using the renter data introduced earlier.

A row-level example

Consider the `Rental Data` data source. Let's say that we know that apartments 1, 2, and 3 are upstairs, while 4, 5, and 6 are downstairs. We'd like to have that information as part of the dataset.

We could potentially add this attribute to the source data, but there are times when this may not be an option. We may not have permission to change the source data or the source might be a spreadsheet that is automatically generated every day and any changes would be overwritten.

Instead, we can create a row-level calculation in Tableau to extend your data. To do so, simply create a calculated field named `Floor` with the code:

```
IF [Apartment] > 3
THEN "Upstairs"
ELSE "Downstairs"
END
```

This code uses an IF THEN ELSE conditional statement and returns a string result. The field shows up in the **Data** window under **Dimensions**. The **Floor** dimension can be used just like any other dimension. It can slice the data, define the level of detail, and group measures.

Row-level calculations can be dimensions, but if the result had been numeric, then Tableau would have placed the field under **Measures** by default. As we've seen before, the default use of a field can be changed from a measure to a dimension, or vice versa by dragging and dropping it within the **Data** window.

Notice that Tableau adds a small equal sign to the icon to indicate it is a calculated field:

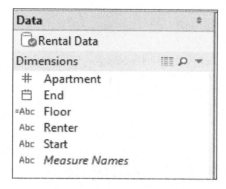

The code is executed for every row of data and returns a value of **Upstairs** if the apartment is greater than 3 or **Downstairs** otherwise. We can verify that the code is operating on a row-level by examining the source data. Simply click on the **View Data** icon next to dimensions to see the row-level detail (it's next to the magnifying glass icon in the preceding screenshot). Here, the new field and row-level values are clearly seen:

A good question to ask yourself whenever you write a calculation in Tableau is, What happens if the data changes? Right now, for example, we only have apartments 1 through 6 in the data. However, what happens if additional apartments are added in future? Or what if bad data shows up tomorrow with an apartment marked as 0? The calculation shown will only work as long as 1 through 3 are the only apartments downstairs and everything else is upstairs. Consider this alternate code:

```
IF [Apartment] >= 1 AND [Apartment] <= 3
  THEN "Downstairs"
ELSEIF [Apartment] > 3 AND [Apartment] <= 6
  THEN "Upstairs"
ELSE "Unknown"
END
```

This code explicitly defines a case for 1 through 3 and 4 through 6 and defaults to Unknown for anything else. When Unknown shows up as a result in data visualization, you will immediately recognize that you have a new apartment.

An aggregate-level example

One thing you might want for analysis is the price per square foot. This does not exist in the data. This really couldn't be stored in the source, because the value changes based on the level of detail (for example, the average price per square foot per month might be different from the average price per square foot per apartment). Rather, it must be calculated at an aggregate level.

Let's create a calculation named Price per Square Foot with the following code:

```
SUM([Price]) / SUM([Area])
```

This code indicates that the sum of Price should be divided by the sum of Area. That is, all values for Price will be added, all values for Area will be added, and then the division will take place.

Once you click on **OK** in the **Calculated Field** dialog box, you'll notice that Tableau places the new field under **Measures**. Tableau will place any calculation with a numeric result under **Measures** by default. However, in this case, there is an additional reason. Tableau will treat every aggregate calculation as a measure—no matter what data type is returned. This is because an aggregate calculation depends on dimensions to define the level of detail at which the calculation is performed. So, an aggregate calculation cannot itself be a dimension. Notice that you are not even able to redefine the new field as a dimension.

Now, create a couple of views to see how the calculation returns different results depending on the level of detail in the view. First, take a look at **Price per Square Foot** by **Floor**:

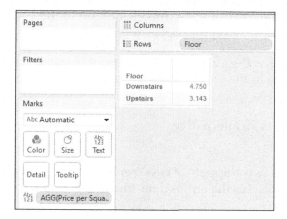

Now, notice how the values change when you add in the **Apartment** field:

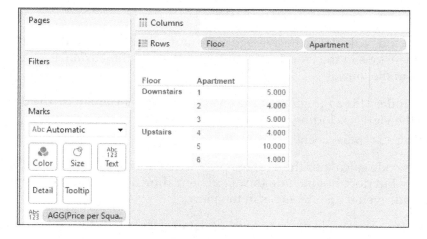

Why did the values change? Because aggregations, including calculated aggregations, depend on which dimensions are defining the level of detail of the view. In the first case, **Floor** was defining the level of detail. So, the calculation added up all the prices for each floor and all the areas for each floor and then divided. In the second case, **Apartment** redefines the level of detail. So, the calculation added up all the prices for each apartment and all the areas for each apartment and then divided.

While nearly every aggregate calculation depends on the dimensions in the view to define the level of detail, **Level of Detail calculations** (introduced in Tableau 9.0) allow you to explicitly define a level of detail that may or may not depend on the dimensions in the view. Consider these variations:

- **Fixed**: This aggregates at the level of detail specified by the list of dimensions in the code regardless of what dimensions are in the view:

```
{FIXED [Floor] : AVG([Price])}
```

  This code returns the average price per floor, regardless of what other dimensions are in the view. Consider the following snippet:

```
{FIXED : AVG([Price])}
```

  The following is an alternative:

```
{AVG([Price])}
```

  Either of these two snippets of code represents a fixed calculation of the average price for the entire data source (or the subset defined by a context filter).

- **Include**: This aggregates at the level of detail determined by the dimensions in the view and the dimensions listed in the code:

```
{INCLUDE [Renter] : AVG([Price])}
```

  This code calculates the average price at the level of detail defined by dimensions in the view but includes the `Renter` dimension, even if `Renter` is not in the view.

- **Exclude**: This aggregates at the level of detail determined by the dimensions in the view, excluding any listed in the code:

```
{EXCLUDE [Apartment] : AVG([Price])}
```

  This code calculates the average price at the level of detail defined in the view but does not include the `Apartment` dimension as part of the level of detail, even if `Apartment` is in the view.

We'll consider an example of using level of detail calculations to solve a complex problem in *Chapter 10, Advanced Techniques, Tips, and Tricks*.

Row level or aggregate – why does it matter?

What if we had created `Price per Square Foot` as a row-level calculation instead of an aggregate-level calculation? Here is the difference in the code:

This is the row-level calculation:

```
[Price] / [Area]
```

This is the aggregate-level calculation:

```
SUM([Price]) / SUM([Area])
```

Here is the difference in the results:

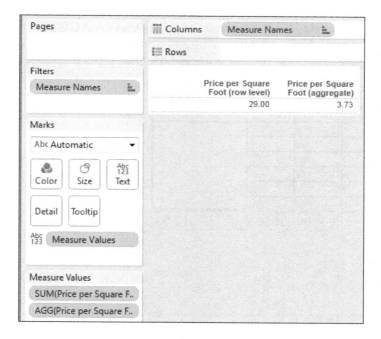

Why is there such a difference in the results? It's the result of the way the calculations are performed.

Notice that the first measure in the preceding view is the sum of `Price per Square Foot (row level)`. That's because the calculation is a row-level calculation, so it gets calculated row by row and then aggregated as a measure after all row-level values have been determined.

The calculation and then final aggregation is performed like this:

Floor	Apartment	Area	Price	Price per Square Foot
Downstairs	1	1000	5000	5
Downstairs	2	500	2000	4
Downstairs	3	500	2500	5
Upstairs	4	1000	4000	4
Upstairs	5	500	5000	10
Upstairs	6	2000	2000	1
				29

Row Level Results: [Price] / [Area]

Sum of Row Level Results

Contrast that with the way the aggregate-level calculation is performed. Notice that the aggregation listed on the active field in the view is **AGG** and not **SUM**. This indicates that you have defined the aggregation in the calculation. Tableau is not further aggregating the results. Here is how the aggregate-level calculation is performed:

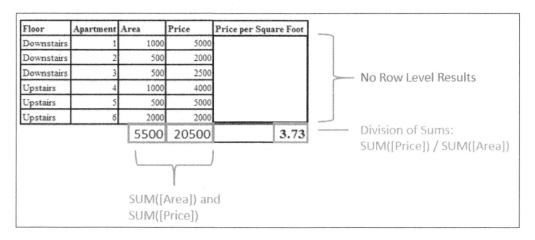

It is vital to understand the difference between row-level and aggregate-level calculations. In general, use row-level calculations when you are certain that the result is something that can be aggregated and make sense at any level of detail. Use aggregate calculations if calculating row-level values would give results that could not then be aggregated correctly.

Parameters

Before moving to some additional examples of row-level and aggregate-level calculations, let's take a little side-trip to examine parameters as they can be used in incredible ways in calculations.

A **parameter** in Tableau is a placeholder for a single, global value such as a number, date, or string. Parameters may be shown as controls (such as sliders, drop-down lists, or type-in textboxes) to end users of dashboards or views, giving them the ability to change the current value of the parameter. The value of a parameter is global so that if the value is changed, every view and calculation in the workbook that references the parameter will use the new value. Parameters provide another way to provide rich interactivity to end users of your dashboards and visualizations.

Parameters can be used to allow anyone interacting with your view or dashboard to dynamically perform the following:

- Alter the results of a calculation
- Change the size of bins
- Change the number of top or bottom items in a top *n* filter or a top *n* set
- Set the value of a reference line or band
- Change the size of bins
- Pass values to a custom SQL statement used in a data source connection

Since parameters can be used in calculations and since calculated fields can be used to define any aspect of a visualization (from filters to colors to rows and columns), the change of a parameter value can have dramatic results. We'll see some examples in the next section.

Creating parameters

To create a parameter, right-click on a field or an empty area in the **Data** window and select **Create Parameter…**. If you right-click on a field and select **Create Parameter…**, Tableau will create a parameter of the same type as the selected field initialized with a list or range of values that match the selected field's values.

The parameter creation interface looks like this:

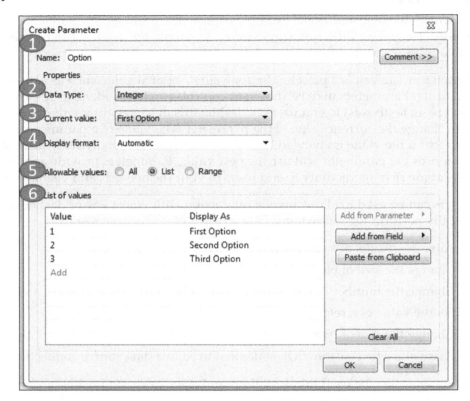

This interface contains the following features:

- **1**: The **Name** field will show as the default title for parameter controls and will also be the reference in calculations.

- **2**: The **Data Type** field defines what type of data is allowed for the value of the parameter. The options include integer, floating point decimal, string, date, or date with time.

- **3**: The **Current value** field defines what the initial default value of the parameter will be. Changing this value in this screen or on a dashboard or visualization where the parameter control is shown will change the current value.

- **4**: The **Display format** field defines how the values will be displayed. For example, we might want to show an integer value as a dollar amount, a decimal as a percentage, or display a date in a specific format.

- **5**: The **Allowable values** option gives us the ability to restrict the scope of values that are permissible. There are three options for **Allowable values**:

  ○ **All**: This allows any input from the user that matches the data type of the parameter.

  ○ **List**: This allows us to define a list of values from which the user must select a single option. The list can be entered manually, pasted from the clipboard, or loaded from a dimension of the same data type. Adding from a field is a one-time operation. If the data changes and new values are added, they will not automatically appear in the parameter list.

  ○ **Range**: This allows us to define a range of possible values, including optional upper and lower limits as well as a step size. This can also be set from a field or another parameter.

- **6**: In the preceding screenshot, the **List of values** field allows us to enter all possible values. In this example, a list of three items has been entered. Notice that the value must match the data type, but the display value can be any string value. This list is static and must be manually updated. Even if you base the parameter on the values present in a field, the list will not change even if new values appear in the data.

If you are using a list of options, consider an integer data type with display values that are easily understood by your end users. The values can be easily referenced in calculations to determine what selection was made, and you can easily change the display value without breaking your calculations. This can also lead to increased performance as comparisons of numeric values are more efficient than string comparisons.

When the parameter is created, it appears in the **Data** window under the **Parameters** section. Right-clicking on a parameter reveals an option to **Show Parameter Control** that adds the parameter control to the view. The little drop-down caret in the upper-right area of the parameter control reveals a menu to customize the appearance and behavior of the parameter control.

Here is the parameter control, shown as **Single Value List**, for the parameter created earlier:

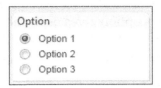

This control can be shown on any sheet or dashboard and allows the end user to select a single value. When the value is changed, any calculations, filters, sets, or bins that use the parameter will be re-evaluated and any views that are affected will be redrawn.

Practical examples of calculations and parameters

Let's turn our attention to some practical examples of calculations. These will be examples of row-level and aggregate-level calculations. These are merely examples. The goal is to learn and understand what is possible with calculations. You will be able to build on these examples as you embark on your analysis and visualization.

Fixing data issues

Often data is not entirely clean. That is, it has problems that need to be corrected before much meaningful analysis can be accomplished. For example, dates may be incorrectly formatted or fields may contain a mix of numeric values and character codes that need to be separated into multiple fields. Calculated fields can often be used to fix these kinds of issues.

We will make use of the sample data introduced at the beginning of this chapter, which contained date values like this:

Start	End
03.17.2014	Oct-14

The original source was Excel and the start date was stored as a text value and the end date was stored as a date. Looking at the workbook of this chapter reveals the following in the **Data** window:

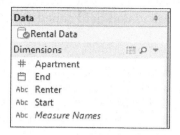

From the icons next to the field names, you'll notice that **Start** has been recognized as a string data type while **End** has been recognized as a date.

When we drop the **Start** field on **Rows**, Tableau does render the value. However, it is rendering it as a text field. There is no date hierarchy or functionality available with the field. If we drop the **End** field on **Rows**, Tableau treats it as a date. However, when we select **Exact Date** and convert it to **Discrete** from the dropdown, we see that the date was stored as the first of the month. **Oct-14** becomes **10/01/2014**.

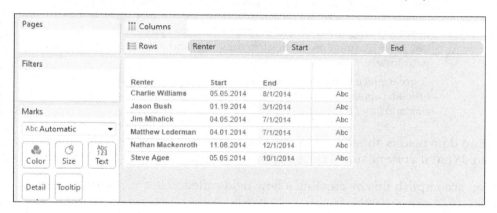

What if we want both **Start** and **End** available as date fields, so we can use all of Tableau's date capabilities, and we also know rentals always run through the end of a month instead of the beginning? We can create calculated fields to solve these issues.

First, let's correctly parse the **Start** date. One thing we might try is changing the data type of the field in Tableau's metadata. If you right-click on the field and go to **Change Data Type | Date**, all the values for **Start** in the view change to Null. This indicates that Tableau truly cannot recognize the date in its current format. So instead, we'll create a new calculated field called Start Date with the code:

```
DATE( REPLACE( [Start], '.', '/' ) )
```

This uses two functions. The innermost function Replace is evaluated first. This replaces every instance of a period (.) in the **Start** field value with a slash (/). A date in this format (for example, 05/05/2014) can be parsed as a date by Tableau. The second function, Date, casts the string value to a date. Since the format can be parsed, the cast function succeeds in giving a valid date. The option to change the data type of calculated fields does not exist as it does for other fields. Instead, we need to cast the result of a calculation to the desired data type in the code.

When we use this new field in the view, Tableau treats it as a date with all the options and formatting normally available for date fields.

> You'll find with calculations that there are often many ways of achieving the same result. Explore the various functions to see whether there are simpler ways of achieving the calculation. For example, the preceding calculation can actually be achieved with the simpler DATEPARSE function using the code DATEPARSE('dd.MM.yyyy', [Start]). The DATEPARSE function takes a format string and a date to return a date with time.
>
> A great place to find help and suggestions for calculations is the official Tableau forums at community.tableausoftware.com/community/forums.

The **End** date field is already recognized as a date, but the value is the first of the month. What if your business rules require it to be the last day of the month?

We can accomplish this by creating a new field called End Date with code such as this:

```
DATEADD('day', -1,
   DATEADD('month', 1, [End])
)
```

This code employs the DATEADD function twice. The innermost function adds a month to the date. This means that April 1, 2014 would become May 1, 2014. Then the outermost function adds -1 day (effectively subtracting a day), so May 1, 2014 becomes April 30, 2014.

We now have two date fields that represent the dates you truly want. The DATEADD function returns a date with time. The default time is midnight. We could use the Date() function to cast the end date as a date without time, or we could set the default formatting of the field to show only the date.

You can use folders in the **Data** window to organize fields. This can be especially useful when building complex calculations requiring multiple fields or calculated fields. The parts can be grouped into folders to make it obvious which fields should be used to build views.

To use folders, right-click on a blank area in the **Data** window and select **Group by Folder**. You can right-click on it again and select **Create Folder**.

Here, for example, the two poorly formatted date fields are placed in a folder:

Fields used in calculations cannot be removed or hidden in a data source connection. However, the folder will remind you not to use the wrong fields in the view and collapsing the folder gets the fields out of sight.

Extending the data

Often there will be dimensions or measures you'd like to have in your data but that are not present in the source. In those cases, you will be able to extend your dataset using calculated fields.

For example, you might want to know the length of each rental contract. You have the start and end dates, but you do not have the length of time between those two dates. Fortunately, this is easy to calculate.

We will now create a calculated field called Length of Contract (days) with the following code:

```
DATEDIFF('day', [Start Date], [End Date])
```

> Tableau employs intelligent code completion. It will offer suggestions for functions and field names as you type in the code editor. Pressing the *Tab* key will autocomplete what you have started to type based on the current suggestion:

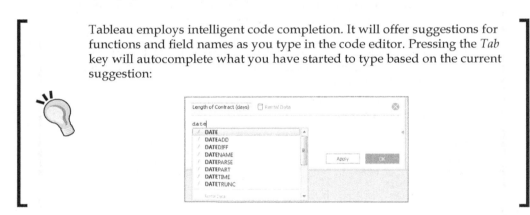

The DATEDIFF function takes a date part description, a start date, and an end date and returns a numeric value for the difference between the two dates. We now have a new measure that wasn't available previously.

We can use the new measure in our visualizations such as the Gantt chart of rentals, as shown in the following screenshot:

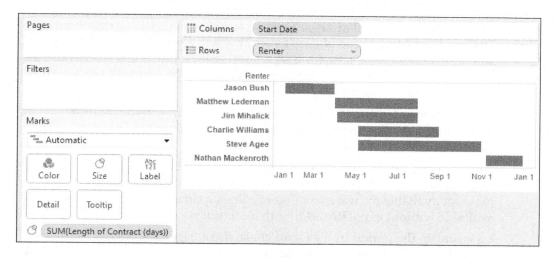

Enhancing user experience, analysis, and visualizations

Calculations and parameters can greatly enhance user experience, analysis, and visualizations.

Let's say we want to give the apartment manager the ability to do some what-if analyses. Every year, they offer a promotional month where any new renter gets a steep discount on rent. This rental manager would love to have a dashboard that gave the ability to pick an arbitrary date and then see how many renters would have gotten the discount.

To accomplish this, follow these steps:

1. Duplicate the Gantt chart from the preceding section.

2. Create a parameter called `Promotional Month Start` with a data type of **Date** and a starting value of `5/1/2014`. This will allow the manager to set and adjust the starting date for the promotional month. Show the parameter control on the view by right-clicking on it in the **Data** window and selecting **Show Parameter Control**.

3. Create a calculated field called `Promotional Month End` that adds a month to the starting month. The code would be `DATEADD('month', 1, [Promotional Month Start])`.

4. Add the `Promotional Month End` field to the level of **Detail** on the **Marks** card. Make sure that the field is set to **Exact Date** and is **Continuous**. This makes it available for use as a reference line or band. Parameters are globally available without explicitly adding them to the view.

5. To visualize the period in the Gantt chart, use a band. Right-click on the date axis in the view and select **Add Reference Line, Band or Box...**. Add a **Band** value for the **Entire Table** option that starts at the parameter start value and ends at the calculated promotional end value. You can set **Label** to none for both the start and the end.

6. Create an additional row-level calculation named `Started in Promotional Period?` that evaluates each start date to determine whether it is in the promotional period. One possibility that simply returns `true` or `false` based on whether the start date falls between the start and end dates is as follows:

    ```
    [Start Date] >= [Promotional Month Start] AND
    [Start Date] < [Promotional Month End]
    ```

7. Place this new calculated field on the **Color** shelf.

We now have a view that allows the apartment manager to change the date and see a dynamically changing view that makes it obvious which renters would have fallen within a given promotional period:

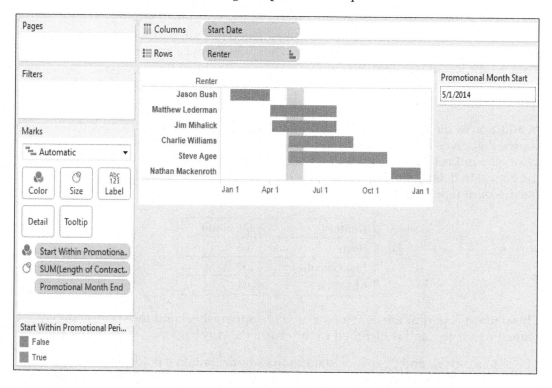

The preceding view shows the proposed promotional month as a band and highlights which rental periods would have started during the month. The band and shading will change as the apartment manager adjusts the promotional start date.

Achieving flexibility with data blends

Data blending, which we examined in detail in *Chapter 2, Working with Data in Tableau*, is a powerful feature with which Tableau can combine two different data sources in a single view. In order for a blend to work, Tableau requires at least one dimension in each dataset that can be linked together (either automatically, when the name and data type match, or manually when you define a relationship by going to **Data | Edit Relationships**). Sometimes, you may not have the required dimensions in one of the datasets or the values may not match. You can use calculated fields as dimensions for blending.

In addition to the renter data we've been using, let's say we have another data source that gives us the promotional discount given to some of the renters. The Discount indicates the percentage by which the rental price is reduced. Ultimately, the goal is to determine the actual price paid by all renters, including the discount if applicable:

Renter	Discount
Bush	.05
Mackenroth	.02
Mihalick	.01

This dataset is part of the Rental Data.xlsx spreadsheet and the connection is named Discount and is included in the chapter's workbook.

The Rental Data and Discount data sources both contain a field called Renter, but the data source with discount only contains the last name of the renter. For the sake of the example, let's assume that all renters, even future ones, have distinct last names. How can we blend these sources together?

Using **Rental Data** as the primary data source and **Discount** as the secondary, this is what happens when we accept the default blending on the **Renter** field:

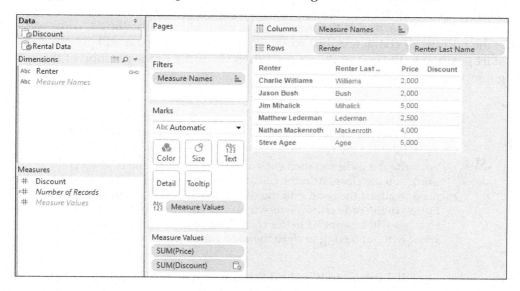

Both **Renter** and **Discount** from the secondary source are NULL. This is because **Renter** in the primary data source is the full name while **Renter** in the secondary source is only the last name.

Data blending occurs on the aliases of fields, so one option is to simply change the aliases for one of the fields to match the values in the other. For example, if you were to right-click on the **Renter** field in the primary source and select **Aliases...**, you could change all the aliases to be the last names only. At that point, the data blend works.

However, if any new renters are added to the data, you'll need to edit the aliases again. That could become tedious or even unmaintainable in a large, constantly changing data source.

Let's consider solving the data blending issue using a calculated field. Create a calculation in the primary data source called Renter Last Name with the code:

```
MID([Renter], FIND([Renter], " ") + 1)
```

This code uses two functions. The first function evaluated is the FIND function that returns the position of the first match of a substring within a string. Here, we're looking for the position of the space in the full name. MID takes a string and then returns a substring starting at the position specified. Optionally, the MID function takes the number of characters, but when we don't supply this, it will give us from the start position to the end of the string. When we add 1 to the position of the space, we get just the last name.

Often you won't be entirely certain what code will get you the exact results the first time. For example, did you know you had to add 1 to the location of the space in the preceding calculation?

So, write the code and then click on **OK** to create the field when you think you are close. Add the calculated field to the view, and if you need to adjust the code, edit the calculated field and use the **Apply** button in the code editor. Every time you click on **Apply**, your code changes will be applied to the view. This allows you to check your work without having to close the code editor window.

Now, we need to tell Tableau to blend on the new calculated field **Renter Last Name**. Manually edit the data relationships (from the menu, go to **Data | Edit Relationships**) and manually match the **Renter Last Name** field from the primary data source to the **Renter** field in the secondary.

At this point, the blend works and we can create a view like this:

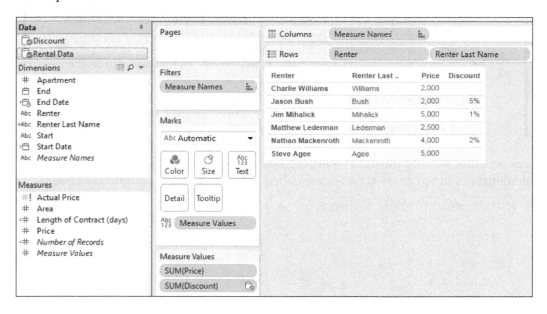

Notice that **Discount** is still blank in several cases. This is because the secondary data source did not have records that matched values from the primary data source. In these cases, the value from the secondary source will be NULL.

We started out with the goal of determining the actual rental price, including any discount, for all renters. We're nearly there. What we'll need to do is create another calculation that takes the initial price and multiplies it by 1 minus the discount. That is, if the discount is .05 (5 percent), we'll multiply the price by .95 (95 percent) to get the actual price.

In order to calculate the actual rental price, we might start with a calculation in the primary source like this:

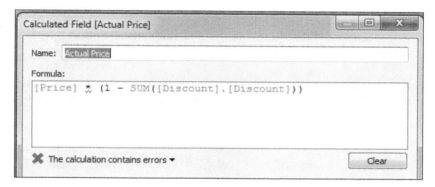

Notice that we can reference fields from the secondary source in calculations. Also notice that you must reference them as aggregates. In this case, we used **SUM** because we knew we'd keep our view at the level of individual renters, and so the sum of the discount would be correct. However, Tableau indicates that the calculation has an error.

The dropdown gives the detail **Cannot mix aggregate and non-aggregate arguments with this function**. This is a common error that indicates that we've tried to mix row-level calculations with aggregate-level calculations. To fix this, we need to identify which elements of the calculation are row level, which are aggregate, and which function is trying to use both. In this case, we are trying to multiply the row-level field **Price** by the aggregate-level field **Discount**. We know **Discount** has to be aggregate since it is from a secondary data source. So, we fix the problem by making **Price** an aggregate.

Our new, valid function looks like this:

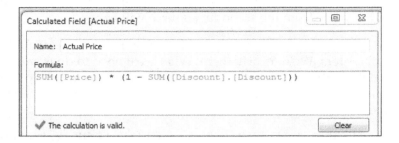

When you drag a field from a secondary source and drop it
into the code editor, Tableau will automatically insert it with
a default aggregation.

However, even though we got a green check mark for valid syntax, our view
indicates that something is still wrong, as shown in the following screenshot:

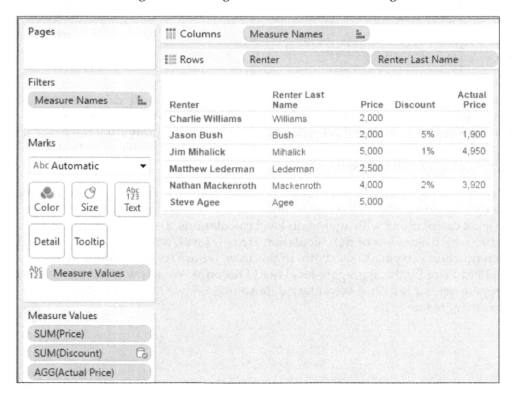

We wanted to get the actual price for all renters, not just the ones that had a match in the secondary data source. The reason we're not seeing what we need is because the calculation returns NULL when **Discount** is NULL. To resolve this, we'll need to adjust the calculation. Edit the code of Actual Price to this:

```
SUM([Price]) * (1 - ZN(SUM([Discount].[Discount])))
```

Here, we've wrapped the aggregation of the secondary **Discount** field in the ZN() function. This function evaluates the expression inside the parentheses and returns 0 if the expression is NULL or simply the expression if it's not NULL. So, in this case, any NULL values for the sum of Discount are converted to 0 and we get the sum of price multiplied by one, which means that we get the original price with no discount.

Renter	Renter Last Name	Price	Discount	Actual Price
Charlie Williams	Williams	2,000		2,000
Jason Bush	Bush	2,000	5%	1,900
Jim Mihalick	Mihalick	5,000	1%	4,950
Matthew Lederman	Lederman	2,500		2,500
Nathan Mackenroth	Mackenroth	4,000	2%	3,920
Steve Agee	Agee	5,000		5,000

The final result is just what we wanted!

Ad hoc calculations

Tableau 9.0 introduced a new way of creating and using calculations. Ad hoc calculations add calculated fields to shelves in a single view without adding fields to the **Data** window.

Let's say you have a simple view that shows the price per renter, like this:

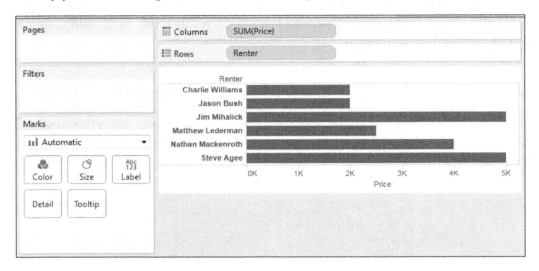

What if you wanted to quickly highlight any renters who had a contract of more than $3,000? One option would be to create an ad hoc calculation. To do so, simply double-click on an empty area of the **Columns, Rows**, or **Measure Values** cards or on the empty space of the **Marks** shelf and then start typing the code for a calculation. In this example, we've double-clicked on the empty space on the **Marks** shelf:

Here, we've entered code that will return `True` if the sum of **Price** is greater than $3,000 and `False` otherwise. Pressing *Enter* or clicking outside the textbox will reveal a new ad hoc field that can be dragged and dropped anywhere within the view. Here, we've added it to the **Color** shelf:

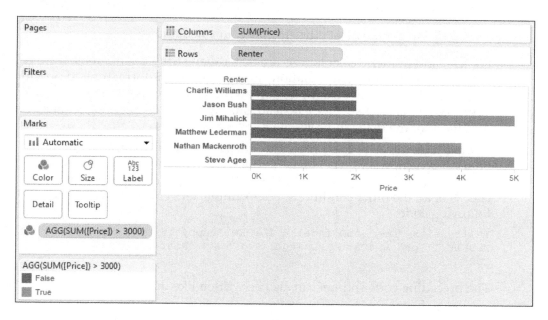

The ad hoc field is only available within the view and does not show up in the **Data** window. You can double-click on the field to edit the code.

 Dragging and dropping an ad hoc field into the **Data** window transforms it into a regular calculated field that will be available for other views using that data source.

Performance considerations

When working with a small dataset and an efficient database, you often won't notice inefficient calculations. With larger datasets, the efficiency of your calculations can start to make a fairly dramatic difference to the speed at which a view is rendered.

Here are some tips to get the most efficiency in your calculations:

- Boolean and numeric calculations are faster than string calculations. If possible, avoid string manipulation and use aliasing or formatting to provide user-friendly labels. For example, don't write this code: `IF [value] == 1 THEN "Yes" ELSE "No" END`. Instead, simply write `[value] == 1`, then edit the aliases of the field, and set `True` to `"Yes"` and `False` to `"No"`.

- Always look for ways to increase the efficiency of a calculation. If you find yourself writing a long `IF`, `ELSEIF` statement with lots of conditions, see whether there are one or two conditions that you can check first to eliminate checks of all the other conditions. For example, let's consider simplifying the following code:

```
IF [Type] = "Dog" AND [Age] < 1 THEN "Puppy"
ELSEIF [Type] = "Cat" AND [Age] < 1 THEN "Kitten"
END
```

  The preceding code snippet can also be written like this:

```
IF [Age] < 1 THEN
    IF [Type] = "Dog" THEN "Puppy"
    ELSEIF [Type] = "Cat" THEN "Kitten"
    END
END
```

  Notice how checking the type doesn't have to be done for any records where the age was less than 1. This could be a very high percentage of records in the dataset.

- Row-level calculations have to be performed for every row of data. Try to minimize the complexity of row-level calculations. However, if that is not possible or doesn't solve a performance issue, consider the next option.

- When you create a data extract, certain row-level calculations are materialized. This means that the calculation is performed once, when the extract is created, and the results are then stored in the extract. This means that the data engine does not have to execute the calculation over and over. Instead, the value is simply read from the extract. Calculations that use any user functions or parameters, or TODAY() or NOW(), will not be materialized in an extract as the value necessarily changes according to the current user, parameter selection, and system time. Tableau's optimizer may also determine not to materialize certain calculations that are more efficiently performed in memory rather than having to read the stored value.

When you use an extract to materialize row-level calculations, only the calculations that were created at the time of the extraction are materialized. If you edit calculated fields or create new ones after creating the extract, you will need to optimize the extract (right-click on the data source connection or select it from the **Data** menu and then go to **Extract | Optimize**).

Summary

Calculations open up amazing possibilities in Tableau. You are no longer confined to the fields in the source data. With calculations, you can extend the data by adding new dimensions and measures, fix bad or poorly formatted data, enhance the user experience with parameters for user input and calculations that enhance the visualizations, and you can achieve flexibility that makes data blending work in situations where the data might have made it difficult or impossible otherwise.

The key to using calculated fields is an understanding of the three levels of calculations in Tableau. Row-level calculations are performed for every row of source data. These calculated fields can be used as dimensions, or they can be further aggregated as measures. Aggregate-level calculations are performed at the level of detail defined by the dimensions present in a view. They are especially helpful, and even necessary, when you must first aggregate components of the calculation before performing additional operations.

In the next chapter, we'll explore the third level of calculations: table calculations. These are some of the most powerful calculations in terms of their ability to solve problems and open up incredible possibilities for in-depth analysis. In practice, they range from very easy to exceptionally complex.

5
Table Calculations

Table calculations are one of the most powerful features in Tableau. They enable solutions that really couldn't be achieved in any other way (short of writing a custom application or complex custom SQL scripts!):

- They make it possible to use data that isn't structured well and still get quick results without waiting for someone to fix the data at the source
- They make it possible to compare and perform calculations on aggregate values across rows of the resulting table
- They open incredible possibilities for analysis and creative approaches to solving problems

Table calculations range in complexity from incredibly easy to create (a couple of clicks) to extremely complex (requiring an understanding of addressing, partitioning, and data densification). We'll start off simple and move toward complexity in this chapter. The goal is to gain a solid foundation to create and use table calculations, understanding how they work, and to see some examples of how they can be used. We'll cover these topics:

- Overview of table calculations
- Quick table calculations
- Scope and direction
- Addressing and partitioning
- Custom table calculations
- Practical examples

Most of the examples here will use the sample `Superstore Sales` data we've used in previous chapters. To follow along with the examples, create a new workbook with a connection to this data.

An overview of table calculations

Table calculations are different from all other calculations in Tableau. Row-level and aggregate calculations, which we considered in the previous chapter, are performed at the data source layer. If you were to examine the queries sent to the data source by Tableau, you'd find the code for your calculations translated into whatever flavor of SQL the data source used.

Table calculations, on the other hand, are performed after the initial query. Here's an extended diagram that shows how aggregated results are stored in Tableau's cache:

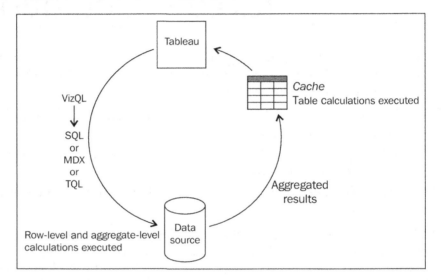

Table calculations are performed on the aggregate table of data in Tableau's cache right before the data visualization is rendered. As we'll see, this is important to understand for multiple reasons, including the following:

- **Aggregation**: Table calculations operate on aggregate data. You cannot reference fields in a table calculation without referencing it as an aggregate.

- **Filtering**: Any filters will be applied before table calculations. This means that table calculations will only be applied to data returned from the source to the cache. You'll need to determine whether you have allowed all the data necessary for table calculations to work as desired.

- **Late filtering**: Filters are applied at the data source level, except for table calculations used as filters. In this case, the row-level and aggregate filters are applied first, the aggregate data is returned to the cache, and then the table calculation is applied as a filter that effectively hides data from the view. This allows for some creative approaches to solving certain kinds of problems that we'll consider in some of the examples.

- **Performance**: If you are using a **Live** connection to an enterprise database server, then row-level and aggregate-level calculations will take advantage of enterprise-level hardware. Table calculations are performed in the cache, which means that they will be performed on whatever machine is running Tableau. You will not likely need to be concerned if your table calculations are operating on a dozen or even hundreds of rows of aggregate data. However, if you are getting back several hundred thousand rows of aggregate data, then you'll need to consider the performance of your table calculations.

Creating and editing table calculations

There are several ways to create table calculations in Tableau:

- Using the drop-down menu for any active field used as a numeric aggregate in the view, select **Quick Table Calculation** and then the desired calculation

- Using the drop-down menu for any active field used as a numeric aggregate in the view, select **Add Table Calculation** and then select the calculation type and adjust any desired settings

- Create a calculated field and use one or more table calculation functions to write your own custom table calculations

The first two options create a quick table calculation that can be edited or removed using the drop-down menu on the field and selecting **Edit Table Calculation...** or **Clear Table Calculation**. The third option creates a calculated field that can be edited or deleted as any other calculated field.

A field on a shelf in the view that is using a table calculation or that is a calculated field using table calculation functions will have a delta symbol icon.

An active field used as numeric aggregate will look like this:

An active field used as a numeric aggregate with table calculation will look like this:

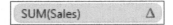

Table calculations can be used in any type of visualization. Most of the examples in this chapter will show them in a text table. As the text table most closely matches the actual aggregate table in the cache, this makes it easier to see how table calculations are working.

When building a view that uses calculations, especially more complex ones, try using a table with all dimensions on the **Rows** shelf and then adding table calculations as discrete values on **Rows** to the right of the dimensions. Once you have all the table calculations working as desired, you can rearrange the fields in the view to give you the appropriate visualization.

Quick table calculations

Quick table calculations are predefined table calculations that can be applied to any field used as a measure in the view. These calculations include common and useful calculations such as the running total, difference, percent difference, percent of total, rank, percentile, moving average, YTD total, compound growth rate, year over year growth, and YTD growth. These calculations do not show up as separate fields in the **Data** window.

Consider the following example using the sample `Superstore Sales` data:

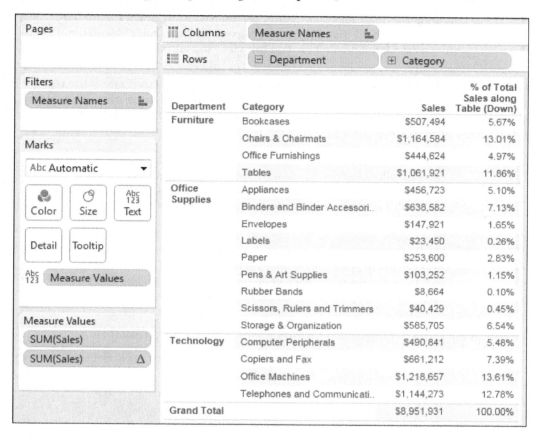

Here, **Sales** is shown twice using **Measure Names** and **Measure Values**. The second **SUM(Sales)** field has the **Percent of Total** quick table calculation applied. Each value shown in the **Sales** column is the sum of sales for that department and category. The value shown in the **% of Total Sales** column is the sum of sales for that department and category divided by the total sum of sales. Using the quick table calculation avoided writing any code.

You can actually see the code that the quick table calculations use by editing the quick table calculation and then clicking on the **Customize...** button in the **Edit** dialog. This will convert the quick table calculation into a calculated field where you can view and edit the code.

Notice that the **Measure Names** field defines column headers that show the names of the measures. For table calculations, the name includes the type of quick table calculation and the scope and direction, in this case **Table (Down)**. We'll take a look at the details of scope and direction shortly.

Tableau will not allow you to have exact duplicate measures on the **Measure Values** shelf. However, Tableau will allow the same measure field with a different aggregation or table calculation to share the **Measure Values** shelf.

The preceding example was created by dragging and dropping the first **Sales** field from the **Data** window onto the pane (which assigned it to the **Text** shelf), applying the quick table calculation, and then dragging and dropping the **Sales** field again on the pane, which allowed it to share the same space using **Measure Names / Measure Values**.

The following table demonstrates the quick table calculations available:

Year of Order Date	Quarter of Order..	Sales	Running Total	Difference	% Difference	% of Total	Rank	Percentile	Moving Average	YTD	Compound Growth Rate	Year over year Growth	YTD Growth
2012	Q1	$415,886.00	$415,886.00			4.65%	12	31.25%	$415,886.00	$415,886.00	0.00%		
	Q2	$352,779.00	$768,665.00	($63,107.00)	-15.17%	3.94%	13	25.00%	$384,332.50	$768,665.00	-15.17%		
	Q3	$456,694.00	$1,225,359.00	$103,915.00	29.46%	5.10%	10	43.75%	$408,453.00	$1,225,359.00	4.79%		
	Q4	$698,986.00	$1,924,345.00	$242,292.00	53.05%	7.81%	5	75.00%	$502,819.67	$1,924,345.00	18.90%		
2013	Q1	$272,065.00	$2,196,410.00	($426,921.00)	-61.08%	3.04%	16	6.25%	$475,915.00	$272,065.00	-10.07%	-34.58%	-34.58%
	Q2	$337,352.00	$2,533,762.00	$65,287.00	24.00%	3.77%	14	18.75%	$436,134.33	$609,417.00	-4.10%	-4.37%	-20.72%
	Q3	$546,388.00	$3,080,150.00	$209,036.00	61.98%	6.10%	6	68.75%	$385,268.33	$1,155,805.00	4.65%	19.64%	-5.68%
	Q4	$788,742.00	$3,868,892.00	$242,354.00	44.36%	8.81%	3	87.50%	$557,494.00	$1,944,547.00	9.57%	12.84%	1.05%
2014	Q1	$294,067.00	$4,162,959.00	($494,675.00)	-62.72%	3.28%	15	12.50%	$543,065.67	$294,067.00	-4.24%	8.09%	8.09%
	Q2	$428,267.00	$4,591,226.00	$134,200.00	45.64%	4.78%	11	37.50%	$503,692.00	$722,334.00	0.33%	26.95%	18.53%
	Q3	$508,189.00	$5,099,415.00	$79,922.00	18.66%	5.68%	9	50.00%	$410,174.33	$1,230,523.00	2.02%	-6.99%	6.48%
	Q4	$1,000,217.00	$6,099,632.00	$492,028.00	96.82%	11.17%	2	93.75%	$645,557.67	$2,230,740.00	6.30%	26.81%	14.72%
2015	Q1	$535,158.00	$6,635,790.00	($464,059.00)	-46.40%	5.99%	7	62.50%	$681,521.33	$535,158.00	2.14%	82.33%	82.33%
	Q2	$518,601.00	$7,154,391.00	($17,557.00)	-3.27%	5.79%	8	56.25%	$684,992.00	$1,054,759.00	1.71%	21.09%	46.02%
	Q3	$722,674.00	$7,877,065.00	$204,073.00	39.35%	8.07%	4	81.25%	$592,477.67	$1,777,433.00	4.03%	42.21%	44.45%
	Q4	$1,074,962.00	$8,952,027.00	$352,288.00	48.75%	12.01%	1	100.00%	$772,079.00	$2,852,395.00	6.54%	7.47%	27.87%

In both cases, the table calculations are being calculated down the table. For example, the running total adds each successive value of the sum of sales until it gets to the bottom of the table. The final row for the running total is equal to the grand total.

You'll notice that some table calculations work in the **Grand Total** line. For example, percent of total is correctly shown as **100%**. Other quick table calculations are blank. For example, **Difference** and **YTD Growth** do not have values in the **Grand Total** line. **Rank** is not blank, but gives a rank of **1** in the grand total. All of these particular table calculations reference more than one row of aggregate data. **Difference**, in this instance, is the difference of one row of aggregate data and the previous row. The **Grand Total** row is a single, separately calculated row. Table calculations computed within the total cannot reference rows outside the total.

The preceding example was created using **Measure Names** and **Measure Values** just as the first example was. The column headers for each table calculation are not the default. For instance, **% of Total** was originally % of Total Sales along Table (Down).

At times, it is helpful to see the full description to understand what the calculation is and how it is being calculated. However, you will at times want to change the headers to make them more user friendly or to save space. To do this, right-click on each column header and select **Edit Alias**.

Scope and direction

Scope and direction are terms that describe how a table calculation is computed relative to the table. When a table calculation is relative, rearranging the fields in the view will not change the scope and direction. Let's see what scope and direction are:

- **Scope**: The scope defines the boundaries within which a given table calculation can reference other values
- **Direction**: The direction defines how the table calculation *moves* within the scope

You've already seen table calculations being calculated using the **Table (Down)** option. In these cases, the scope was the entire table and the direction was down. For example, the **Running Total** calculation ran from top to bottom of the entire table, adding values as it *moved*.

To define scope and direction for a table calculation, use the drop-down menu for the field in the view and select **Compute using**. You will get a list of options that will vary slightly depending on the dimensions present in the view. The first of the options listed allows you to define the scope and direction relative to the table. After the option for **Cell**, you will see a list of dimensions present in the view. We'll take a look at those in the next section.

Options for scope and direction relative to the table are as follows:

- **Scope options**: **Table**, **Pane**, and **Cell**
- **Direction options**: **Down**, **Across**, **Down then Across**, and **Across then Down**

In order to understand these options, consider the following example:

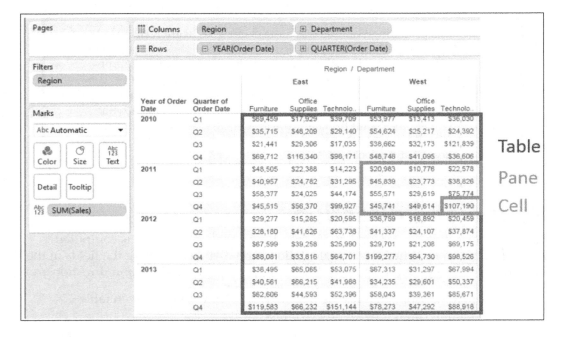

As you can see, Tableau considers all the aggregate data to be part of the table. The pane is often defined by the second to last dimension on the **Rows** and/or **Columns** shelf. In the preceding screenshot, you can see that the intersection of **YEAR** on **Rows** and **Region** on **Columns** defines the panes (one is highlighted, but there are actually eight in the view). The cell is defined by the intersection of all dimensions in the view (one is highlighted, but there are actually 96 in the view).

Working with scope and direction

In order to see how scope and direction work together, let's work through a few examples. We'll start by creating our own custom table calculation. Create a new calculated field named `Index` with the `Index()` code.

 `Index` is a table calculation function that starts with the value 1 and increments by one as it moves along a given direction and within a given scope. There are many practical uses for `Index`, but we'll use it here because it is easy to see how it is moving for a given scope and direction.

Create the table shown in the previous example with **YEAR(Order Date)** and **QUARTER(Order Date)** on **Rows** and **Region** and **Department** on **Columns**. Instead of placing **Sales** in the view, add the newly created **Index** field to the **Text** shelf. Then, experiment using the drop-down menu on the **Index** field and select **Compute using** to cycle through various scope and direction combinations. In the following examples, we've only kept the **East** and **West** regions and the years **2010** and **2011**:

- **Table (Across)**: By default, Tableau uses **Table (Across)**. In the following screenshot, notice how **Index** increments across the entire table:

Year of Order Date	Quarter of Order Date	East			West		
		Furnitu..	Office Supplies	Techno..	Furnitu..	Office Supplies	Techno..
2010	Q1	1	2	3	4	5	6
	Q2	1	2	3	4	5	6
	Q3	1	2	3	4	5	6
	Q4	1	2	3	4	5	6
2011	Q1	1	2	3	4	5	6
	Q2	1	2	3	4	5	6
	Q3	1	2	3	4	5	6
	Q4	1	2	3	4	5	6

- **Table (Down)**: When using **Table (Down)**, **Index** increments down the entire table:

Year of Order Date	Quarter of Order Date	East			West		
		Furnitu..	Office Supplies	Techno..	Furnitu..	Office Supplies	Techno..
2010	Q1	1	1	1	1	1	1
	Q2	2	2	2	2	2	2
	Q3	3	3	3	3	3	3
	Q4	4	4	4	4	4	4
2011	Q1	5	5	5	5	5	5
	Q2	6	6	6	6	6	6
	Q3	7	7	7	7	7	7
	Q4	8	8	8	8	8	8

- **Table (Across then Down)**: This increments **Index** across the table, then steps down and continues to increment across; this repeats for the entire table:

Year of Order Date	Quarter of Order Date	East			West		
		Furnitu..	Office Supplies	Techno..	Furnitu..	Office Supplies	Techno..
2010	Q1	1	2	3	4	5	6
	Q2	7	8	9	10	11	12
	Q3	13	14	15	16	17	18
	Q4	19	20	21	22	23	24
2011	Q1	25	26	27	28	29	30
	Q2	31	32	33	34	35	36
	Q3	37	38	39	40	41	42
	Q4	43	44	45	46	47	48

- **Pane (Across)**: This defines a boundary for **Index** and causes **Index** to increment across the table until it reaches the pane boundary at which point the indexing restarts:

Year of Order Date	Quarter of Order Date	East			West		
		Furnitu..	Office Supplies	Techno..	Furnitu..	Office Supplies	Techno..
2010	Q1	1	5	9	1	5	9
	Q2	2	6	10	2	6	10
	Q3	3	7	11	3	7	11
	Q4	4	8	12	4	8	12
2011	Q1	1	5	9	1	5	9
	Q2	2	6	10	2	6	10
	Q3	3	7	11	3	7	11
	Q4	4	8	12	4	8	12

- **Pane (Down then Across)**: This allows **Index** to increment down the pane and continue by stepping across the table. The pane defines the boundary here:

Year of Order Date	Quarter of Order Date	East			West		
		Furnitu..	Office Supplies	Techno..	Furnitu..	Office Supplies	Techno..
2010	Q1	1	2	3	1	2	3
	Q2	1	2	3	1	2	3
	Q3	1	2	3	1	2	3
	Q4	1	2	3	1	2	3
2011	Q1	1	2	3	1	2	3
	Q2	1	2	3	1	2	3
	Q3	1	2	3	1	2	3
	Q4	1	2	3	1	2	3

Scope and direction work with any table calculation. Consider how a running total or percent difference would be calculated using the same movement and boundaries shown in the examples. The examples in the list don't cover every possible combination of scope and direction relative to the table. Keep experimenting with different options until you feel comfortable with how scope and direction work.

Addressing and partitioning

Addressing and partitioning are very similar to scope and direction but are most often used to describe how table calculations are computed with absolute reference to certain fields in the view. With addressing and partitioning, you define which dimensions in the view define the partition (scope) and which dimensions define the addressing (direction). Using addressing and partitioning gives you much finer control because your table calculations are no longer relative to the table layout and you have many more options for fine-tuning the scope, direction, and order of the calculations.

To begin to understand how this works, let's consider a simple example. Using the view from the preceding example, set the **Compute using** value of the **Index** field to the **Department** dimension.

Year of Order Date	Quarter of Order Date	East			West		
		Furniture	Office Supplies	Technology	Furniture	Office Supplies	Technology
2012	Q1	1	2	3	4	5	6
	Q2	1	2	3	4	5	6
	Q3	1	2	3	4	5	6
	Q4	1	2	3	4	5	6
2013	Q1	1	2	3	4	5	6
	Q2	1	2	3	4	5	6
	Q3	1	2	3	4	5	6
	Q4	1	2	3	4	5	6

What this does is tell Tableau to compute **Index** along (in the direction of) the selected dimension. In other words, you have used **Department** for addressing. All other dimensions in the view are implicitly used for partitioning, that is, they define the scope or boundaries at which the Index function must restart.

The preceding view looks identical to what you would see if you set **Index** to compute using **Pane (Across)**. However, there is a major difference. When you use **Pane (Across)**, **Index** is always computed across the pane, even if you rearrange the dimensions in the view, remove some, or add others. However, when you compute using a dimension for addressing, the table calculation will always compute using that dimension. Removing that dimension will break the table calculation, and you'll need to adjust the **Compute using** option. If you rearrange dimensions in the view, **Index** will continue to be computed along the **Department** dimension.

Here, for example, is the result of clicking on the **Swap Rows and Columns** button in the toolbar:

Region	Department	2010				2011			
		Q1	Q2	Q3	Q4	Q1	Q2	Q3	Q4
East	Furniture	1	1	1	1	1	1	1	1
	Office Supplies	2	2	2	2	2	2	2	2
	Technology	3	3	3	3	3	3	3	3
West	Furniture	1	1	1	1	1	1	1	1
	Office Supplies	2	2	2	2	2	2	2	2
	Technology	3	3	3	3	3	3	3	3

Notice that **Index** continues to be computed along **Department** even though the entire orientation of the table has changed. (If you've been following the examples in Tableau, swap the rows and columns back before following the next examples).

Advanced addressing and partitioning

In addition to quickly setting the addressing to a single dimension, Tableau allows you to explicitly define addressing and partitioning. Using the drop-down menu on the **Index** field in the view, select **Edit Table Calculation...**. You'll see a dialog box that looks similar to this:

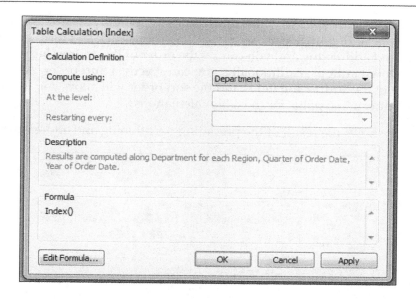

Observe how the options in the **Compute using** dropdown on this screen nearly match the **Compute using** options on the field drop-down menu. You can select any relative scope and direction and use any single dimension for addressing. At the bottom of the drop-down options, there is one additional option: **Advanced...**. Selecting this option reveals another dialog box like this:

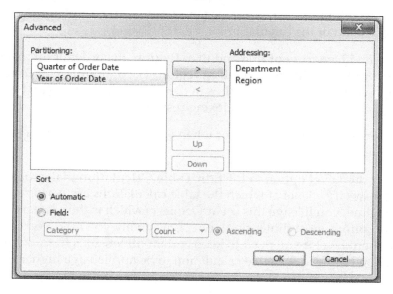

Here you can explicitly specify which dimensions define the partitioning and which define the addressing. You can move dimensions back and forth between partitioning and addressing and you can set the order of the addressing fields using the **Up** and **Down** buttons. Additionally, you can specify a **Sort** order. As we will see, the order of the addressing fields and the sort order will allow you to specify very specific directional paths for the table calculations.

The following shows the result of the preceding partitioning and addressing:

Year of Order Date	Quarter of Order Date	East			West		
		Furniture	Office Supplies	Technol..	Furniture	Office Supplies	Technol..
2010	Q1	1	3	5	2	4	6
	Q2	1	3	5	2	4	6
	Q3	1	3	5	2	4	6
	Q4	1	3	5	2	4	6
2011	Q1	1	3	5	2	4	6
	Q2	1	3	5	2	4	6
	Q3	1	3	5	2	4	6
	Q4	1	3	5	2	4	6

As was specified, **Year of Order Date** and **Quarter of Order Date** are defining the partition. This means that each unique combination of year and quarter defines one partition (the first is outlined in the screenshot). You can see that addressing has defined the direction of the calculation.

Since we specified **Department** first and then **Region**, the order of indexing starts with the **Furniture** department in the **East** region (**1**), continuing with **Furniture** in the **West** region (**2**), then **Office Supplies** in the **East** region (**3**), **Office Supplies** in the **West** region (**4**), and so on. The indexing restarts every partition.

There are a few other things to consider when working with addressing and partitioning:

- The **At the Level** option in the **Edit Table Calculation** dialog box allows you to specify a level at which the table calculations are performed. Most of the time, you'll leave this set at **Deepest** (which is the same as setting it to the bottom-most dimension), but occasionally you might want to set it at a different level if you need to keep certain dimensions from defining the partition but need the table calculation to be applied at a higher level.

- The **Restarting Every...** option effectively makes the field selected and all dimensions in the addressing above that field selected part of the partition, but allows you to maintain the fine-tuning of the ordering.

- Dimensions are the only kinds of fields that can be used in addressing, however a discrete (blue) measure can be used to partition table calculations. To enable this, use the drop-down menu on the field and uncheck **Ignore in Table Calculations**.

Advanced table calculations

Before we move on to some practical examples, let's briefly consider advanced table calculations. Advanced, in this case, simply means that code is written instead of using a **Quick Table Calculation** option. You can see a list of available table calculation functions by creating a new calculation and selecting **Table Calculation** from the dropdown under **Functions**.

You can think of table calculations broken down into several categories. In each of the examples, we'll go to **Compute using | Category**, which means **Department** is the partition. The various advanced table calculations include:

- **Meta-table functions**: These are functions that give you information about the partitioning and addressing. These functions also include Index, First, Last, and Size.

Department	Category	Index	First	Last	Size
Furniture	Bookcases	1	0	3	4
	Chairs & Chairmats	2	-1	2	4
	Office Furnishings	3	-2	1	4
	Tables	4	-3	0	4
Office Supplies	Appliances	1	0	8	9
	Binders and Binder Accessori..	2	-1	7	9
	Envelopes	3	-2	6	9
	Labels	4	-3	5	9
	Paper	5	-4	4	9
	Pens & Art Supplies	6	-5	3	9
	Rubber Bands	7	-6	2	9
	Scissors, Rulers and Trimmers	8	-7	1	9
	Storage & Organization	9	-8	0	9
Technology	Computer Peripherals	1	0	3	4
	Copiers and Fax	2	-1	2	4
	Office Machines	3	-2	1	4
	Telephones and Communicati..	4	-3	0	4

First gives the offset from the first row in the partition. So, the first row in each partition is 0. **Last** gives the offset to the last row in the partition. **Size** gives the size of the partition. **Index, First,** and **Last** are all affected by scope/partition and direction/addressing, while **Size** will give the same result at each address of the partition no matter what direction is specified.

- **Lookup and previous value:** The first of these two functions gives you the ability to reference values in other rows while the second gives you the ability to carry forward values. Notice that direction is very important for these two functions:

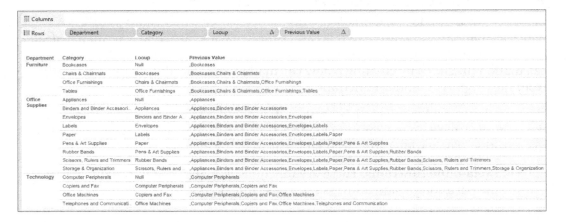

Here, we've used the `Lookup(ATTR([Category]), -1)` code that looks up the value of **Category** in the row offset by `-1` from the current. The first row in each partition gets a null result from the lookup (because there isn't a row before it).

For `Previous Value`, we used the following code:

```
Previous_Value("") +
"," +
ATTR([Category])
```

Notice that in the first row, there is no previous value, so `Previous_Value` simply returned what we specified as the default: an empty string. This was then concatenated together with a comma and the category in that row, giving us the value, **Bookcases**.

In the second row, **Bookcases** is the previous value that gets concatenated with a comma and the category in that row, giving us the value, **Bookcases, Chairs & Chairmats**, which becomes the previous value in the next row. The pattern continues throughout the partition and then restarts in the partition defined by the **Office Supplies** department.

- **Running functions**: These functions run along the direction/addressing and include `Running_Avg`, `Running_Count`, `Running_Sum`, `Running_Min`, and `Running_Max`.

Department	Category	Sales	Running Sum	Running Min
Furniture	Bookcases	$507,494	507,494	507,494
	Chairs & Chairmats	$1,164,584	1,672,079	507,494
	Office Furnishings	$444,624	2,116,703	444,624
	Tables	$1,061,921	3,178,624	444,624
Office Supplies	Appliances	$456,723	456,723	456,723
	Binders and Binder Accessori..	$638,582	1,095,305	456,723
	Envelopes	$147,921	1,243,226	147,921
	Labels	$23,450	1,266,676	23,450
	Paper	$253,600	1,520,276	23,450
	Pens & Art Supplies	$103,252	1,623,528	23,450
	Rubber Bands	$8,664	1,632,192	8,664
	Scissors, Rulers and Trimmers	$40,429	1,672,621	8,664
	Storage & Organization	$585,705	2,258,326	8,664
Technology	Computer Peripherals	$490,841	490,841	490,841
	Copiers and Fax	$661,212	1,152,052	490,841
	Office Machines	$1,218,657	2,370,709	490,841
	Telephones and Communicati..	$1,144,273	3,514,982	490,841

Notice that `Running_Sum(SUM[Sales]))` continues to add the sum of sales to a running total for every row in the partition. `Running_Min` keeps the value of the sum of sales if it is the smallest value it has encountered so far as it moves along the rows of the partition.

- **Window functions**: These functions operate across all rows in the partition at once and essentially aggregate the aggregates. They include `Window_Sum`, `Window_Avg`, `Window_Max`, `Window_Min`, and others.

Department	Category	Sales	Window Sum	Window Max
Furniture	Bookcases	$507,494	3,178,624	1,164,584
	Chairs & Chairmats	$1,164,584	3,178,624	1,164,584
	Office Furnishings	$444,624	3,178,624	1,164,584
	Tables	$1,061,921	3,178,624	1,164,584
Office Supplies	Appliances	$456,723	2,258,326	638,582
	Binders and Binder Accessori..	$638,582	2,258,326	638,582
	Envelopes	$147,921	2,258,326	638,582
	Labels	$23,450	2,258,326	638,582
	Paper	$253,600	2,258,326	638,582
	Pens & Art Supplies	$103,252	2,258,326	638,582
	Rubber Bands	$8,664	2,258,326	638,582
	Scissors, Rulers and Trimmers	$40,429	2,258,326	638,582
	Storage & Organization	$585,705	2,258,326	638,582
Technology	Computer Peripherals	$490,841	3,514,982	1,218,657
	Copiers and Fax	$661,212	3,514,982	1,218,657
	Office Machines	$1,218,657	3,514,982	1,218,657
	Telephones and Communicati..	$1,144,273	3,514,982	1,218,657

- **Rank functions**: These functions provide various ways to rank based on aggregate values.

Department	Category	Sales	Rank
Furniture	Bookcases	$507,494	3
	Chairs & Chairmats	$1,164,584	1
	Office Furnishings	$444,624	4
	Tables	$1,061,921	2
Office Supplies	Appliances	$456,723	3
	Binders and Binder Accessori..	$638,582	1
	Envelopes	$147,921	5
	Labels	$23,450	8
	Paper	$253,600	4
	Pens & Art Supplies	$103,252	6
	Rubber Bands	$8,664	9
	Scissors, Rulers and Trimmers	$40,429	7
	Storage & Organization	$585,705	2
Technology	Computer Peripherals	$490,841	4
	Copiers and Fax	$661,212	3
	Office Machines	$1,218,657	1
	Telephones and Communicati..	$1,144,273	2

- **R script functions**: These functions allow integration with R, an analytics platform that can be used for complex scripting.

- **Total**: The `Total` function deserves its own category because it functions a little differently from the others. Unlike the other functions that work on the aggregate table in the cache, `Total` will requery the underlying source for all the source data rows that make up a given partition. In many cases, this will yield the same result as a `window` function.

  For example, `Total(SUM([Sales]))` gives the same result as `Window_Sum(SUM([Sales]))`, but `Total(AVG([Sales]))` will likely give a different result from `Window_AVG(SUM([Sales]))` because `Total` is giving you the actual average of underlying rows while the `window` function is averaging the sums.

Practical examples

Having looked at some of the foundational concepts of table calculations, let's consider some practical examples. We'll start with some simple ones and move toward complexity.

Moving Average

We've been looking at table calculations mostly in text-tables because this is a natural way to think about how they work. It's also a great way to build and verify table calculations prior to transforming the view into a more complex visualization.

The following is a view of daily website traffic for the `www.VizPainter.com` blog. Observe that the top time series displays a cyclical up-down pattern. What causes this? It turns out that most people visit the blog during the week. Very few individuals read it on weekends.

Seeing this pattern can be useful, but it can also be distracting if we really just care about the overall trends and pattern.

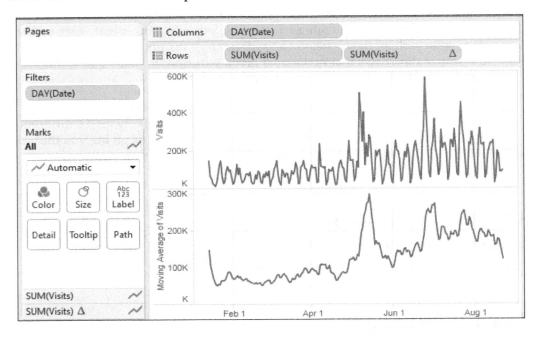

To "smooth out" the chart, the second **Visits** field on the **Rows** shelf uses the **Moving Average** quick table calculation to "smooth out" the chart with **Moving Average** applied. The options were edited to show each day as the moving average of the previous 7 days, moving **Table (Across)**. This was accomplished by editing the table calculation using the drop-down menu on the field.

Be careful when using rolling averages so that you don't distort the data story. Notice in the preceding view that the overall pattern is easier to see when the cyclic pattern was smoothed.

However, certain peaks and valleys are also smoothed. In fact, the highest point on the top line chart is the second highest on the bottom. In some ways, **Moving Average** highlights the importance of the first peak (it lasted longer), but it may also obscure other insights. Often, it is helpful to show both lines (side by side, as in the preceding screenshot, or using a dual axis).

Ranking within higher levels

You may run across something like this when you attempt to sort:

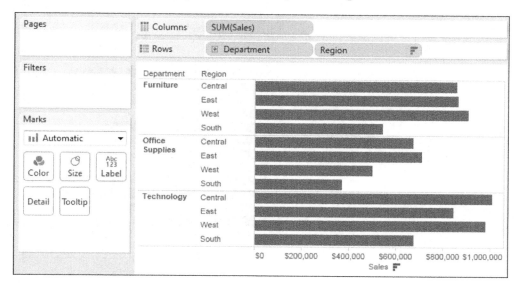

In this case, **Region** has been sorted in descending order by the sum of sales. Notice that the order is an overall ordering. The **Central** region is always the first within each department because it is the highest in overall sales. However, it is not the highest in each **Department**. For example, it is third for **Furniture**.

If you wanted to see a different sort order for each department, you might consider an approach like the following. First, create a field named `Rank` with the code `Rank()`.

Placing that field on **Rows** just before **Region** causes Tableau to sort by **Department** first, **Rank** second, and then **Region**. Continuous fields will always be shifted to the right of discrete (blue) fields on **Rows** or **Columns**, so you'll need to change **Rank** to discrete to place it before **Region**. Set the **Compute using** option to **Pane (Down)** or **Region** to achieve a rank within the department.

The final view has the row headers for **Rank** hidden (use the drop-down field on **Rank** to uncheck **Show Row Headers**) and a copy of **Rank** on the **Label** shelf:

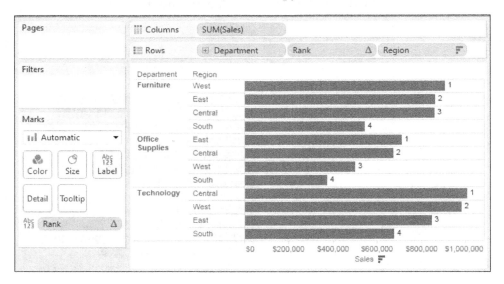

Late filtering

Let's say you've built a view that allows you to see the percent age of total sales for each department. You have already used a quick table calculation on the **Sales** field to give you a percent age of the total. You've also used **Department** as a quick filter. However, this presents a problem.

Since table calculations are performed after the aggregate data is returned to the cache, the filter on **Department** has already been evaluated at the data source and the aggregate rows don't include any departments excluded by the filter. Thus, the percent of total will always add up to 100 percent; that is, it is the percent age of the filtered total.

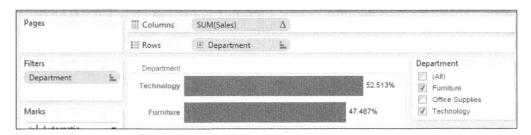

What if you wanted to see the percent age of the total sales for all departments, even if you want to exclude some from the display? One option is to use a table calculation as a filter.

If you create a calculated field called Department (late filter) with the LOOKUP(ATTR([Department]), 0) code and place that on the **Filters** shelf instead of the **Department** dimension, then the filter is not applied at the source, the aggregate data is visible to other table calculations, and the table calculation filter merely "hides" departments from the final view, as shown here:

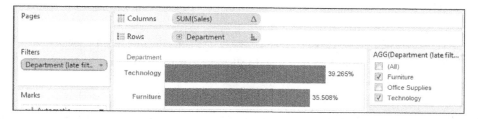

You might have noticed the ATTR function used. Remember that table calculations require aggregate arguments. ATTR (short for attribute) is a special aggregation that returns the value of a field if there is only a single value of that field present for a given level of detail or * if there is more than one value.

To understand this, experiment with a view that has both **Department** and **Category** on **Rows**. Using the drop-down menu on the active field in the view, change **Category** to **Attribute**. It will display as * because there is more than one category for each department. Then, undo and change **Department** to **Attribute**. It will display the department name because there is only one department per category.

Last occurrence

Let's say you want to know the sales amount for your top three customers the last time they made an order. This is fairly easy to accomplish with the following steps:

1. We'll set up the example by filtering the data to the top three customers. Accomplish this by dragging and dropping the **Customer Name** field to the **Filters** shelf, selecting the **Top** tab, and setting it to the top **3** by **Sum** of **Sales**. Remember that this filter will occur before any table calculations are applied.

2. Build a simple view with **Customer Name** on **Rows** first and then **Order Date** as a continuous (green) **Exact Date**. The default view is a Gantt chart, but we have not specified a length for the Gantt bars.

3. Create a calculated field called `Last` with the `Last()` code and place it on the **Text** shelf. Notice that the default computation is **Table (Across)** and counts down to 0 as the last order date for each customer.

4. Move the **Last** field from **Text** to the **Filters** shelf and keep only a range of `0` to `0`. This will eliminate all marks from the view except the last order date for each customer.

> The `First()` and `Last()` functions are also quite useful when working with data that has duplicate records. By filtering to keep only the first or last row in the view, you can eliminate unwanted values.

5. Drag and drop **Sales** from the **Data** window to the **Text** shelf. You now have the sales amount for the last order placed by each customer.

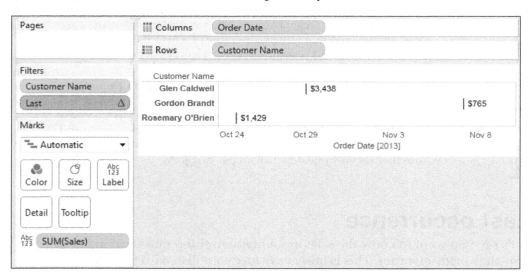

At times, you may experience an unexpected behavior in Tableau. Try following the steps mentioned again, but this time, in step 2, change **Order Date** from continuous (green) to discrete (blue). When you place the **Last** field on the **Text** shelf in step 3, you'll see that each customer has the same values of **Last** for every date, even dates where they did not have an order. If you continue to work through the example, you'll reach a result in which every customer is shown to have the same "last" order date.

What's going on? The combination of certain table calculations and dimensions on the **Rows** and **Columns** shelves (specifically, in this case, the Last() function and presence of a discrete date) causes a relatively undocumented behavior known as **data densification**.

Data densification refers to any time when rows are added by Tableau to the aggregate table in the cache. It is an umbrella term that covers domain padding (where missing values in dates or bins are filled in) and domain completion (where missing combinations of dimensions are completed by adding rows of aggregate data).

In this case, the aggregate table in the cache has been filled out to include all possible combinations of customers and dates. At times, this can be quite useful, but here it is not at all what we want. One way to turn off data densification is to move the dimension(s) causing the data densification to the level of **Detail** on the **Marks** card and replacing them on **Columns** or **Rows** with the attribute.

In the example, you could copy (by pressing *Ctrl* and dragging and dropping) the **Order Date** field from **Rows** to the level of **Detail** on the **Marks** card. Then, using the drop-down menu on the **Order Date** field on **Rows**, change it to **Attribute**.

Joe Mako and Jonathan Drummey are among those who have provided clarity on the topic of data densification. A search of key terms related to data densification will yield multiple sources of information on the topic.

Summary

We've covered a lot of concepts surrounding table calculations in this chapter. You now have a solid foundation to understand everything from quick table calculations to advanced table calculations. The practical examples we covered barely scratch the surface of what is possible, but should give you an idea of what can be achieved. The kind of problems that can be solved and the diversity of questions that can be answered are almost limitless!

We'll turn our attention to some lighter topics such as formatting and design in the next couple of chapters, but we'll certainly see another table calculation or two before we're finished!

6
Formatting a Visualization to Look Great and Work Well

Formatting is about more than just making data visualization look good. Presentation can make a huge difference in the way it is received and understood. As you move beyond making great discoveries and performing great analysis, you'll want to consider how you will present the story of the data.

Tableau's formatting options give you quite a bit of flexibility. Fonts, titles, captions, colors, row and column banding, labels, shading, annotations, and much more can all be customized to make your visualizations tell a story well.

This chapter will cover the following topics:

- Formatting considerations
- How formatting works in Tableau
- Adding value to visualizations

Formatting considerations

Tableau employs good practices for formatting and visualization from the time you start dropping fields on shelves. You'll find that the discrete palettes use colors that are easy to distinguish, fonts are generally acceptable, grid lines are faint, and numbers and dates follow the default format settings defined in the metadata.

The default formatting is certainly adequate for discovery and analysis. If you are focused on analysis, you may not want to spend too much time fine-tuning the formatting until you have moved on in the cycle. However, when you start to consider how you will communicate data to others, you will need to contemplate how adjustments to the formatting can make a difference in how well the data story is told.

 Sometimes, you will have certain formatting preferences in mind when you start your design. In these cases, you might set formatting options in a blank workbook and save it as a template.

Here are some of the things you should consider:

- **Audience**: Who is the audience and what are their requirements?

- **Setting**: This is the environment in which the data story is communicated. Is it a formal business meeting where the format should reflect a high level of professionalism? Is it going to be shared on a blog to informally, or playfully tell a story?

- **Mode**: How will the visualizations be presented? You'll want to make sure rows, columns, fonts, and marks are large enough for a projector or compact enough for an iPad. If you are publishing to Tableau Server, Tableau Online, or Tableau Public, then did you select fonts that are safe for the web?

- **Mood**: Certain colors, fonts, and layouts elicit different emotional responses. Does the data tell a story that should invoke a certain response from your audience? The color red, for example, may connote danger, negative results, or indicate that an action is required. However, you'll need to be sensitive to your audience and the specific context. Colors have different meanings for different cultures. Red might not be a good choice to communicate negativity if it is also the color of the corporate logo.

- **Consistency**: Generally, use the same fonts, colors, shapes, line thickness, and row banding throughout all visualizations. This is especially true when they will be seen together in a dashboard or even used in the same workbook. You may also consider how to remain consistent throughout the organization without being too rigid.

All of these considerations will inform your formatting decisions. As with everything else you do with Tableau, think of formatting as an iterative process. Look for feedback from your intended audience often and adjust as necessary to make sure your communication is as clear and effective as possible. The goal of formatting is to more effectively communicate the data.

How formatting works in Tableau

Tableau uses default formatting that includes default fonts, colors, shading, and alignment. Additionally, there are several levels of formatting you can customize:

- **Workbook level**: The following type of formatting comes under this category:

    ○ **Default field formatting**: Using the drop-down menu on any field in the **Data** window, go to **Default Properties | Date Format** or **Default Properties | Number Format**. This sets the default format in Tableau's metadata and will be applied to any view where custom formatting has not been applied.

- **Story level**: When viewing a story, go to **Format | Story** (or **Story | Format**) to edit formatting for story-specific elements.

- **Dashboard level**: Dashboard-specific elements can be formatted. When viewing a dashboard, go to **Format | Dashboard** (or **Dashboard | Format**) to specify the formatting for dashboard titles, subtitles, shading, and text objects.

- **Worksheet level**: We'll consider the various options. The following types of formatting are available for a worksheet:

    ○ **Sheet formatting**: This formatting includes font, alignment, shading, borders, and lines.

    ○ **Field-level formatting.** This formatting includes fonts, alignment, shading, and number and date formats. This formatting is specific to how a field is displayed in the current view. The options you set at a field level override defaults set at a worksheet level. Number and date formats will also override the default field formatting.

    ○ **Additional formatting.** Additional formatting can be applied to titles, captions, tooltips, labels, annotations, reference lines, field labels, dashboards, stories, and more.

- **Rich text formatting**: Titles, captions, annotations, labels, and tooltips all contain text that can be formatted with varying fonts, colors, and alignment. This formatting is specific to the individual text element.

Worksheet-level formatting

You've already seen how to edit metadata in previous chapters, and we'll cover dashboards and stories in detail in future chapters. So, let's start by considering worksheet-level formatting. Worksheet-level formatting is accomplished using the **Format** window, which appears on the left, in place of the **Data** window.

To view the **Format** window, select **Format** from the menu and then **Font**, **Alignment**, **Shading**, **Border**, or **Lines**.

 You can also right-click on nearly any element in the view and select **Format**. This will open the **Format** window specific to the context of the element you selected. Just be certain to verify the title of the **Format** window matches what you expect. When you make a change, you should see the view update immediately to reflect your formatting. If you don't, you are likely to be working in the wrong tab of the formatting window.

You should now see the **Format** window on the left. It will look similar to this:

Notice these key aspects of the formatting window:

- The title of the window will give you the context for your formatting selections. Always pay attention to the title.

- The icons on the top match the selection options in the **Format** menu. This allows you to easily navigate through these options without returning to the menu each time.

- The three tabs, **Sheet**, **Rows**, and **Columns**, allow you to specify options at a sheet level and then override these options and defaults at a row and column level.

- The **Fields** dropdown in the upper-right corner allows you to fine-tune formatting at a field level.

- Any changes that you make will be previewed and will result in a bold label to indicate that the formatting option has been changed from the default.

Here are the results of the preceding formatting settings, in a view with the **Grand Total** column and all subtotals turned on (enable totals from the menu by going to **Analysis | Totals**):

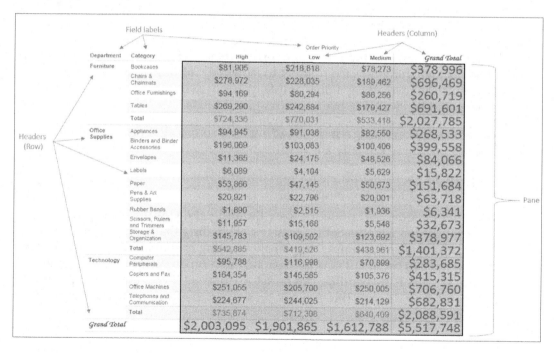

The preceding view demonstrates various parts of the view as they relate to formatting:

- The labels of **Department**, **Category**, and **Order Priority**, are **field labels**.

- The distinct values for each field defining a header (for example, **Furniture**, **Office Supplies**, **Bookcases**, **High**, **Low**, and so on) make up **headers**. Row headers are on the left, and column headers are on top.

- A faint outline and shading have been added to the view to highlight the **pane**. You'll recall a slightly different definition of a pane when we discussed table calculations. Here, when talking about formatting, pane refers to the window defined by headers and axes.

Field labels for rows or columns can be hidden by right-clicking on a field label and selecting the **Hide Field Labels** option. You can unhide field labels by using the menu, that is, by going to **Analysis | Table Layout** and selecting the appropriate option.

Compare the **Format** window shown previously with the resulting view. Notice how the default settings are broken down by **Pane**, **Header**, and **Tooltip**. **Total** and **Grand Total** are also broken down by **Pane** and **Header**. This allows you to set a default, sheet-level, formatting for headers and text that appear in the pane (as a text mark or label) and then further override these defaults for subtotals and grand totals. By switching tabs from **Sheet** to **Rows** or **Columns**, you are able to further fine-tune the formatting. For example, you can make row grand totals have different pane and header fonts from column grand totals.

The other format options (**Alignment**, **Shading**, and so on) all work very similarly to the **Font** option. There are only a few subtleties to mention:

- **Alignment** includes options for horizontal and vertical alignment, text direction, and text wrapping.

- **Shading** includes an option for **Row Banding** and **Column banding**. Banding allows for alternating patterns of shading that help to differentiate or group rows and columns. Light row banding is enabled by default for text tables, but can be useful in other visualization types such as horizontal bar charts as well. Row banding can be set to different levels that correspond to the number of discrete (blue) fields present on the **Rows** or **Columns** shelf.

- **Borders** refers to borders drawn around cells, panes, and headers. It includes options for **Row Divider** and **Column Divider**. You can see, in the preceding view, the dividers between the departments. By default, the level of the borders is set based on the next field to the last field on **Rows** or **Columns**.

If all of your fields on **Rows** and **Columns** are dimensions (or discrete aggregates with the **Ignore in Table calculations** option unchecked), then the borders match up nicely with the definition of pane used for table calculations. This gives you a good way of visually seeing the pane.

For example, notice the row and column dividers (formatted for emphasis) in the following view clearly defining four panes:

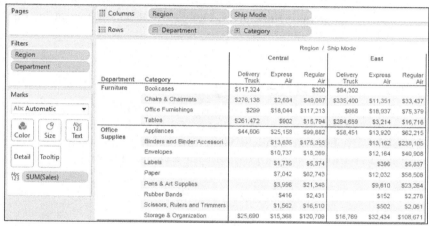

- **Lines** refers to lines that are drawn on visualizations using an axis. This includes grid lines, reference lines, zero lines, and axis rulers. You can access a more complete set of options for reference lines and drop lines from the **Format** option on the menu.

Clearing formatting

The three options for clearing the format are as follows:

- **Clear single option**: In the **Format** window, right-click on the title or control of any single option you have changed and select **Clear** from the pop-up menu.

- **Clear all current options**: At the bottom of the **Format** window, click on the **Clear** button to clear all visible changes. This applies only to what you are currently seeing in the **Format** window. For example, if you are looking at **Shading** and the **Rows** tab and then click on **Clear**, only the shading options you have changed on the **Rows** tab will be cleared.

- **Clear sheet**: From the menu, go to **Worksheet | Clear | Formatting**. You can also use the dropdown from the clear item on the toolbar. This clears all custom formatting on the current worksheet.

Field-level formatting

In the upper-right corner of the **Format** window is a little dropdown labeled **Fields**. Selecting this dropdown gives you a list of fields in the current view and selecting a field updates the **Format** window with options appropriate for the field. Here, for example, is the window as it appears for the **SUM(Sales)** field:

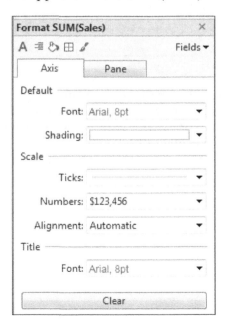

The title of the **Format** window will alert you to the field you are formatting. Selecting an icon for font, alignment, and so on will switch back to sheet-level formatting. However, you can switch between the tabs: **Axis** (for continuous fields) or **Header** (for discrete fields) and **Pane**. The options for fields include font, alignment, shading, and number and date formats. The latter two options will override any default metadata formats.

Custom number formatting

When you alter the format of a number, you can select from several standard formats as well as a custom format. The custom format allows you to enter a format string that Tableau will use to format the number. The format string allows up to three entries, separated by semicolons, to represent the positive, negative, and zero formats. Here are some examples assuming the positive number 34,331.336 and the negative number -8,156.7777:

Format string	Result
#;-#	**34331** and **-8157**
#,###.##;(#,###.##)	**34,331.34** and **(8,156.78)**
#,###.000000; -#,###.000000	**34,331.336000** and **-8,156.777700**
"up" #,###;"down" #,###;"same"	**up 34,331** and **down 8,157**

Notice how, when the custom format string does not have enough precision, Tableau rounds the display of the number. Always be aware that numbers you see as text, labels, or headers may have been rounded due to the format.

Also, observe how you can mix format characters such as the pound sign, commas, and decimal points with strings. The final example here would give a label of `"same"` where 0 would normally have been displayed.

Selecting a predefined format that is close to what you want and then switching to custom will allow you to start with a custom format string that is close.

An additional aspect of formatting a field is specially formatting NULL values. When formatting a field, select the **Pane** tab and locate the **Special Values** section:

Enter any text you would like to display in the pane (in the **Text** field) when the value of the field is null. You can also choose where marks should be displayed. The **Marks** dropdown gives multiple options that define where and how the marks for null values should be drawn when an axis is being used. You have the following options:

- **Show at Indicator**: This results in a small indicator with the number of null values in the lower-right corner of the view. You can click on the indicator for options to filter the nulls or show them at the default value. You can right-click on the indicator to hide it.

- **Show at Default Value**: This displays a mark at the default location (usually 0).

- **Hide (Connect Lines)**: This does not place a mark for null values, but does connect lines between all non-null values.

- **Hide (Break Lines)**: This causes the line to break where there are "gaps" created by not showing the null values.

You can see these options in the following screenshot:

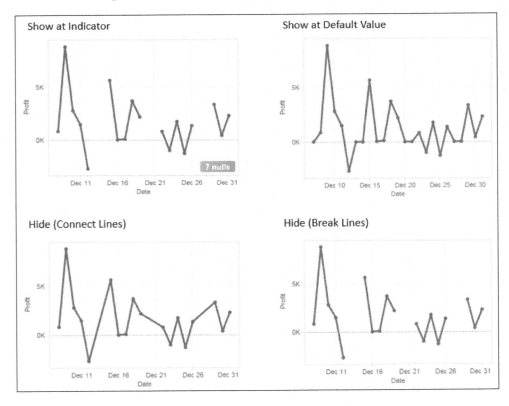

You'll notice the line charts here have little circle markers at the location of each mark drawn in the view. When the mark type is a line, clicking on the **Color** shelf opens a menu that gives options for the markers. All mark types have standard options such as **Color** and **Transparency**. Some mark types support additional options such as **Border** and/or **Halo**:

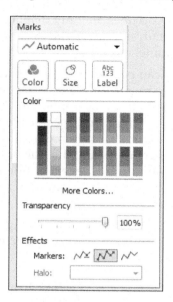

Additional formatting options

Additional formatting options can also be accessed from the formatting window. These options include:

- A myriad of options for **Reference Lines**
- Line and text options for **Drop Lines**
- Shading and border options for **Titles and Captions**
- Text, box, and line options for **Annotations**
- Font, shading, alignment, and separator options for **Field Labels**
- Title and body options for **Legends**, **Quick Filters**, and **Parameters**
- **Cell Size** and **Workbook Theme** options

You'll find most of these fairly straightforward. A few options might not be as obvious:

- Drop lines, which appear as lines drawn from the mark to the axis, can be enabled by right-clicking on any blank area in the pane of the view with an axis and going to **Drop Lines | Show Drop Lines**. Additional options can be accessed using the same right-click menu and selecting **Edit Drop Lines**. Drop lines are only displayed in Tableau Desktop and Reader but are not currently available when a view is published to Tableau Server, Online, or Public.

- Titles and captions can be shown or hidden for any view by selecting **Worksheet** from the menu and then the desired options. In addition to standard formatting that can be applied to titles and captions, the text of a title or caption can be edited and specifically formatted by double-clicking on the title or caption, right-clicking on the title or caption and selecting **Edit**, or by using the drop-down menu of the title or caption (or the drop-down menu of the view on a dashboard). The text of titles and captions can dynamically include the values of parameters, values of any field in the view, and certain other data and worksheet-specific values.

- Annotations can be created by right-clicking on a mark or space in the view and selecting **Annotate** and then selecting one of the following three types of annotations:

  ○ **Mark**: These annotations are associated with a specific mark in the view. If that mark does not display (due to a filter or an axis range), then neither will the annotation. Mark annotations can include a display of the values of any fields that define the mark or its location.

  ○ **Point**: These annotations are anchored to a specific point in the view. If the point is ever not visible in the view, the annotation will disappear. Point annotations can include a display of any field values that define the location of the point (for example, the coordinates of the axis).

  ○ **Area**: These annotations are contained within a rectangular area. The text of all annotations can dynamically include the values of parameters, certain other data, and worksheet-specific values.

You can copy formatting from one worksheet to another (within the same workbook or across workbooks) by selecting **Copy Formatting** from the **Format** menu while viewing the source worksheet (or select the **Copy Formatting** option from the right-click menu on the source worksheet tab). Then, select **Paste Formatting** on the **Format** menu while viewing the target worksheet (or select this option from the right-click menu on the target worksheet tab).

This option will apply any custom formatting present on the source sheet to the target. However, specific formatting applied during the editing of the text of titles, captions, labels, and tooltips is not copied to the target sheet.

Adding value to visualizations

Now that we've considered how formatting works in Tableau, let's take a look at some ways in which formatting can add value to a visualization.

When you apply custom formatting, always ask yourself what the formatting adds to the understanding of the data. Is it making the visualization clearer and easier to understand? Or is it just adding clutter and noise?

In general, try a minimalistic approach. Remove everything from the visualization that isn't necessary. Emphasize important values, text, and marks while de-emphasizing those that are only providing support or context.

Consider the following visualization:

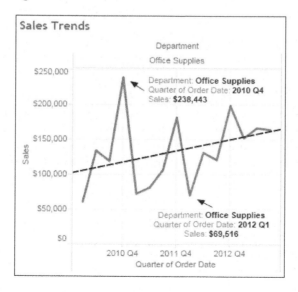

Now let's consider this visualization:

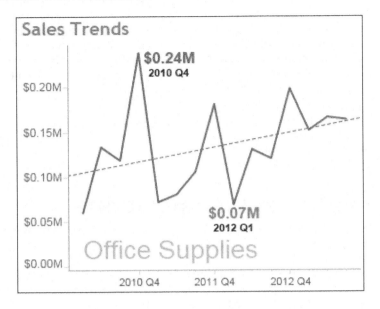

Both the preceding charts are showing sales by quarter, filtered to the **Office Supplies** department. With the exception that the first view has the **Department** field on **Columns** in an attempt to make it clear that only office supplies sales are being shown, the field arrangement for the two views is exactly the same. The first view uses the default formatting.

Consider some of the customizations in the second view:

- The size and color of the title have been adjusted, making it larger but slightly fainter. It is a little easier to read without demanding quite as much immediate attention.

- The **Sales** field has been formatted so it uses a custom currency with two decimal places and units of millions. This is true for the axis and the annotations. Often a high level of precision can clutter visualization. The initial view of the data gives the trend and enough detail to understand the order of magnitude. Tooltips or additional views can be used to reveal detail and precision.

- The axis labels have been removed by right-clicking on the axis, selecting **Edit Axis** and then clearing the text. The title of the view clearly indicates that one is looking at **Sales**. The values alone reveal the second axis to be by quarter. If there are multiple dates in the data, you might need to specify which one is in use.

- The gridlines have been removed. Gridlines can add value to a view, especially in views where being able to determine values is of high importance. However, they can also clutter and distract. You'll need to decide, based on the view itself and the story you are trying to tell, whether gridlines are helpful or not.

- The thickness of line has been decreased by clicking on the **Size** shelf and adjusting the size.

- The trend line has been formatted to have a lighter weight and match the color of the line.

- The lines and arrows have been removed from the annotations to reduce clutter.

- The size and color of the annotations have been altered to make them stand out. If the goal had been to simply highlight the minimum and maximum values on the line, labels might have been a better choice as they can be set to display at only **Min/Max**. In this case, however, the lower number is actually the second lowest point in the view.

- A third annotation (an area annotation) has been added to the lower-left corner of the view to make it clear that the chart is for **Office Supplies**. The font is large but light, so it is clear but not loud. Alternatively, this information could have been added to the title, which may have kept the chart even cleaner.

- Row and column borders have been removed.

- Axis rulers have been added to compensate for the lack of row and column dividers (axis rulers are available under the **Lines** option on the **Format** window).

Formatting can also be used to dramatically alter the appearance of visualizations. Consider the following chart:

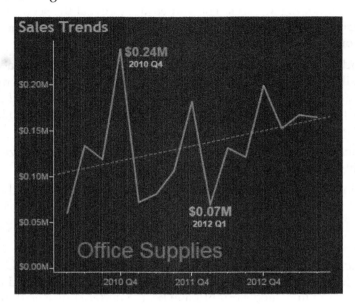

This visualization is nearly identical to the previous view. However, shading has been applied to the worksheet and the title. Additionally, fonts were lightened or darkened as needed to show up well on a black background. Some find this format more pleasing, especially on mobile devices. If the view is to be embedded in a website with a dark theme, this formatting may be very desirable. However, you may find some text more difficult to read on a dark background. You'll want to consider your audience, the setting, and mode of delivery as you consider whether such a format is the best for your situation.

Sequential color palettes (a single color gradient based on a continuous field) should be reversed when using a black background. This is because the default of lighter (lower) to darker (higher) shade works well on a white background where darker colors stand out and lighter colors fade into white.

On a black background, lighter colors stand out more and darker colors fade into black. You'll find the reverse option when you edit a color palette using the drop-down menu on the legend, double-clicking on the legend, or right-clicking on the legend, then selecting **Edit Colors...**, and checking **Reversed**.

Tooltips

As they are not always visible, tooltips are an easily overlooked aspect of visualizations. However, they add a subtle professionalism. Consider the default tooltip that displays when the end user hovers over one of the marks in the preceding view:

Compare the default tooltip to this tooltip:

The tooltip was edited using the menu option **Worksheet** | **Tooltip**, which brought up an editor, allowing the rich editing of text in the tooltip:

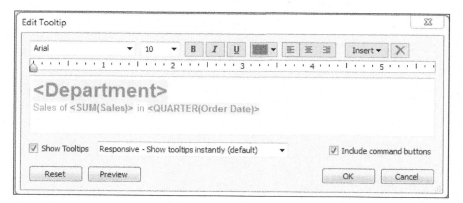

This editor is similar to those used to edit the text of labels, titles, captions, and annotations. Observe the **Insert** dropdown in the upper-right corner that allows you to insert fields, parameters, and other dynamic values. In the text, these are enclosed as a tag. You'll notice that the first tooltip in this screenshot included command buttons (keep and exclude filters, creating sets, groups, and viewing the data). You'll need to decide whether the command buttons add easy functionality or are distracting or confusing for your intended audience. The functionality is always available to the end user via right-click.

[You can keep a tooltip from showing at all by clearing the text. Alternatively, uncheck the **Show Tooltips** option in the editor.]

Summary

The goal of formatting is to increase the effective communication of data. Always consider the audience, setting, mode, mood, and consistency as you work through the iterative process of formatting. Look for formatting that adds value to your visualization and avoid useless clutter. With an understanding of how formatting works in Tableau, you'll have the ability to refine the visualizations you created in discovery and analysis into an incredibly effective communication of your data story. We'll look next at how it all comes together in dashboards.

7
Telling a Data Story
with Dashboards

As you engage in data discovery and analysis, you will create numerous data visualizations. Each of these visualizations gives a snapshot of a story within the data. Each view into the data answers one or maybe a couple of questions. At times, the discovery and analysis are enough to make a key decision, and the cycle is complete. Many times, however, you will need to bring these visualizations together so that they communicate a comprehensive story to your intended audience.

Tableau allows you to bring together related data visualizations into a single dashboard. This dashboard could be a static view of various aspects of the data or a fully interactive environment that allows users to dynamically filter, drill down, and interact with data visualizations.

This chapter will cover the following topics:

- Objectives of dashboards
- Examples of dashboards
- Interactivity with actions
- Story points

We'll take a look at most of these concepts in the context of several in-depth examples where we'll walk through the dashboard design process. Along the way, you'll see some step-by-step instructions. As has been the case before, don't worry about memorizing lists of instructions. Instead, focus on understanding why and how components and aspects of dashboards work.

For the examples, we'll use the Superstore Sales sample data we've used in previous chapters. The .twbx workbook of this chapter in the Learning Tableau\ Workbooks directory contains completed examples. If you'd like to follow along and build from scratch, create a new workbook with a connection to Learning Tableau\Data\Superstore.tde.

Dashboard objectives

Every dashboard seeks to tell a story by giving a clear picture of a certain set of information. Before designing a dashboard, you should understand what story the data tells. How you tell the story will depend on numerous factors, such as your audience, the way the audience will access the dashboard, and what response you want to elicit from your audience.

Stephen Few defines a dashboard as a "visual display of the most important information needed to achieve one or more objectives; consolidated and arranged on a single screen so the information can be monitored at a glance." This definition is helpful because it places some key boundaries around the data story and the way we will seek to tell it in Tableau. In general, your story should follow these guidelines:

- The story must focus on the most important information. Anything that does not communicate or support the main story should be excluded. You may include this information in other dashboards.

- The story that you tell must meet your key objectives. Your objectives may range from giving information, to providing an interface for further exploration, to prompting your audience to take action or make key decisions.

- The main story should be easily accessible and the primary idea should be obvious.

From a Tableau perspective, a dashboard is a set of worksheets along with other various components (such as legends, quick filters, parameters, text, containers, images, and web objects) arranged on a single canvas. Ideally, the visualizations and components should work together to tell a complete and compelling data story. Dashboards are usually interactive.

When you set out to build a dashboard, you'll want to carefully consider your objectives. Your discovery and analysis should have uncovered various insights into the data and its story. Now, it's your responsibility to package this discovery and analysis into a meaningful communication of the story to your particular audience, in a way that meets your objectives and their needs.

Here are some possible approaches to building dashboards based on your objectives. It is by no means a comprehensive list:

- **Guided analysis**: You've done the analysis, made the discoveries, and thus have a deep understanding of the implications of the data story. Often, it can be helpful to design a dashboard that guides your audience through a similar process of making the discoveries for themselves so the need to act is clear. For example, you may have discovered wasteful spending in the manufacturing department but the finance team may not be ready to accept your results unless they can see how the data led you to that conclusion.

- **Exploratory**: Many times, you do not know what story the data will tell when the data is refreshed in the next hour, week, or year. What may not be a significant aspect of the story today might be a major decision point in the future. In these cases, your goal is to provide your audience with an analytical tool that gives them the ability to explore and interact with various aspects of the data on their own. For example, today customer satisfaction is high across all products. However, your dashboard needs to give the marketing team the ability to continually track satisfaction over time, dynamically filter by region and price, and observe any correlations with quality.

- **Scorecard / status snapshot**: There may be wide agreement on KPIs or metrics that indicate good versus poor performance. You don't need to guide the audience through discovery or force them to explore. They just need a top-level summary and enough detail and drilldown to quickly find and fix problems and reward success. For example, you may have a dashboard that simply shows how many support tickets are still unresolved from the previous week. A snapshot image of the dashboard sent to the manager's e-mail via subscription every morning might be enough to spur action.

- **Narrative**: This type of dashboard tells a clear story. There may be aspects of exploration, guided analysis, or performance indication, but primarily, you are showing what is necessary to communicate the meaning of the data. For example, you may desire to tell the story of the outbreak of a disease including where, when, and how it spread. Your dashboard tells the story using the data in a visual way.

We'll take a look at several in-depth examples to better understand a few of these different approaches. Along the way, we'll incorporate many of the skills we've covered in previous chapters, and we'll introduce key aspects of designing dashboards in Tableau.

Example – is least profitable always unprofitable?

Let's say you've been tasked with helping management for a superstore chain to find which items are the least profitable. The management feels that most of the least profitable items should be eliminated from the inventory. However, as you've done your analysis, you've discovered that certain items, while not profitable overall, have made profit at times in various locations. Your primary objective is to lead management through the discovery of the least profitable items and then guide them through a simple analysis that identifies whether an item has always been unprofitable.

Building the views

Let's start by creating the individual views that will comprise your dashboard:

1. Create a bar chart showing the profit by category. Sort the categories in a descending order by the sum of the profit.

2. Add the **Department** field to **Filters** and show a quick filter. You can do this by right-clicking on the **Department** field and selecting **Show Quick Filter** or using the drop-down menu on the field in the view and selecting **Show Quick Filter**.

3. Name the sheet `Profit by Category`.

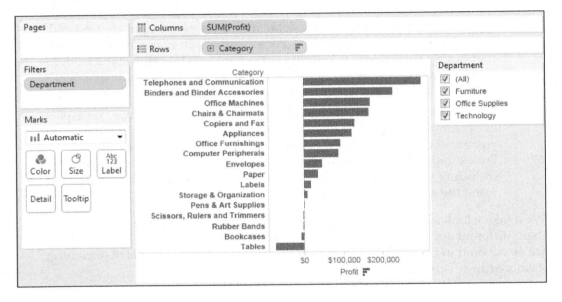

4. Create another, similar view, showing the profit by the item. Sort the items in descending order by the sum of the profit.

5. You'll notice that there are too many items to see at one time. For your objectives on this dashboard, you can limit the items to the top 10 least profitable ones. Add the **Item** field to the **Filters** shelf, select the **Top** tab, and adjust the settings to filter **By field**. Specify the bottom 10 by **Sum(Profit)**.

6. Rename the sheet Top 10 Least Profitable Items.

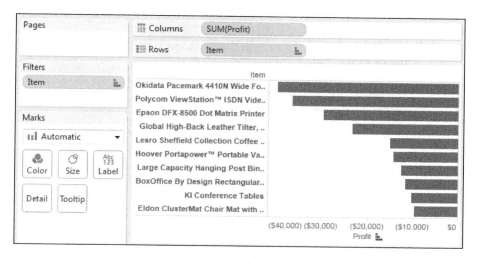

7. Create another sheet that displays a filled map of the profit by state. You can accomplish this rather quickly by double-clicking on the **State** field in the **Data** window and then dropping **Profit** on the **Color** shelf.

8. Rename the sheet to Where?.

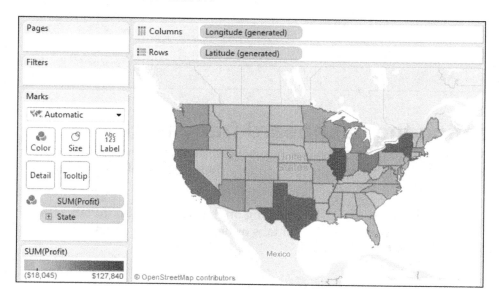

9. Create one final sheet to show when profits were made or lost. Note that the **Order Date** field has been added as the **Quarter** date value and is continuous (green).

10. Turn on mark labels by clicking on the button on the toolbar or explicitly adding **Profit** to the **Label** shelf.

11. Rename the sheet to When?.

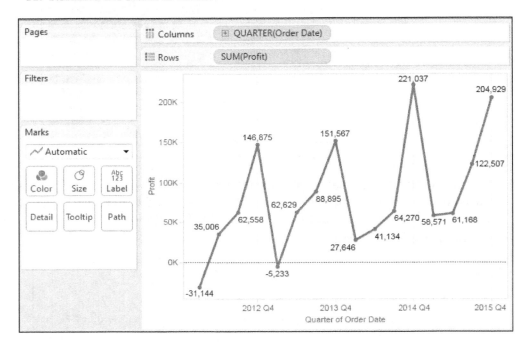

Creating the dashboard framework

At this point, you have all the views necessary to achieve the objectives for your dashboard. Now, all that remains is to arrange them and enable the interactivity required to effectively tell the story:

1. Create a new dashboard by clicking on the **New Dashboard** tab to the right of all existing worksheet tabs or by navigating to **Dashboard | New Dashboard** from the menu.

2. Add the views to the dashboard and arrange them as shown here:

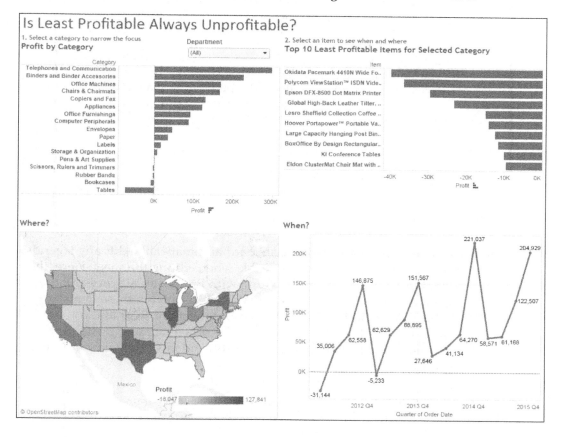

3. Use a **Text** object to give the dashboard the `Is Least Profitable Always Unprofitable?` title.

Every object and container in a dashboard will display a border and several key controls when selected. The grip, in the middle of the top border, allows you to drag and drop the object to another location on the dashboard. The down caret opens a drop-down menu that gives you various options. For example, you can change the appearance and behavior of a quick filter, format an object, or hide titles and captions for a view. The **X** control will remove the object from the dashboard:

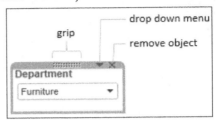

4. Notice how adding a view to the dashboard automatically adds any legends, quick filters, or parameters that are visible when editing the sheet. Here, the **Department** quick filter was added along with **Profit by Category** and the color legend for **Profit** was added with **Where?** By default, these items are added to the right. Select the color legend and the quick filter and convert them to floating objects (either using the drop-down menu or by holding *Shift* while moving them with the grip control) and position them as shown in the preceding screenshot.

Many times, you will need to include a legend, parameter, or quick filter that was not originally shown on a view in the dashboard. To add these after a view has been added, simply use the drop-down caret menu to select the object you wish to add.

5. Note that the **Department** filter is only applied to the **Profit by Category** sheet. Although it will not ultimately be necessary for this dashboard, go ahead and apply the filter to all worksheets on the dashboard. You can accomplish this using the drop-down menu on the quick filter and navigating to **Apply to Worksheets | Selected Worksheets...** and then checking all worksheets. Also note that changing the selection of the quick filter now updates the entire dashboard.

6. Using the same drop-down menu, adjust the quick filer control to show as **Multiple Value (Dropdown)**.

7. Using the drop-down menu, navigate to **Profit by Category** to **Fit | Entire View**. This will ensure all categories are visible without the need for a scrollbar.

8. Edit the title text for **Profit by Category** and **Top 10 Least Profitable Items** to include some instructions for the user, as shown. You can edit the text of titles, captions, and text objects by double-clicking on the text or using the drop-down menu.

Implementing actions to tell the story

You now have a framework that will support the telling of the data story. Your audience will be able to locate least profitable items within the context of a selected category. Then, the selection of an item will answer the question as to whether it has always been unprofitable in every location. To enable this flow and meet your objectives, you'll often need to enable interactivity. In this case, we'll use actions. We'll start by finishing the example with some specific steps and then unpacking the intricacies of actions:

1. Using the menu, navigate to **Dashboard | Actions....** On the resulting dialog, click on **Add Action** and select **Filter...** as the type of action.

2. In the **Add Filter Action** dialog, set **Source Sheets** to **Profit by Category** and **Target Sheets** to **Top 10 Least Profitable Items for Selected Category**. Set the action to run on **Select** and set the option to clear the selection to **Exclude all values**. Name the filter `Filter by Category` and then click on **OK** on this dialog, and then click on the **Actions** dialog.

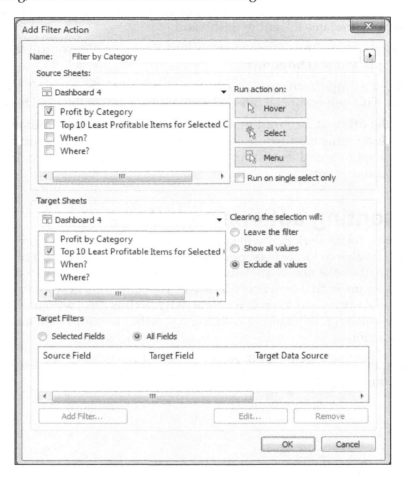

What you now have is an action that is run whenever you select one or more marks in the **Profit by Category** view. The action results in a filter being set on the top 10 view. Based on the option you selected, when you clear the selection the filter will exclude all values on the target, resulting in a blank view.

Note that clicking on a bar in the **Profit by Category** view results in a filtered view of the top 10 items. When you click on the same bar or a blank area in the view, the bar is deselected and the top 10 items view has all values excluded, resulting in a blank view.

You may have noticed that when you select a single item, you have less than 10 items in the top 10 view. For example, selecting **Tables** results in only three items being shown. This is because the top item filter is evaluated at the same time as the action filter. There are only three items with the category of **Tables** that are also in the top 10.

What if you want to see the top 10 items within the category of **Tables**? You can accomplish this using **context filters**.

Context filters are a special kind of filter in Tableau; they are applied before other filters. Other filters are then applied within the "context" of context filters.

Context filters are used by Tableau to create a temporary table (in the source database, if possible; otherwise, in memory) of the subset of data resulting from the context filters. Subsequent queries are then run against the temporary table. With higher volumes of data, this can result in a performance hit when the temporary table is first constructed (and whenever it is reconstructed based on a context filter change) with potentially better subsequent performance because only a subset of data is being queried.

In this case, navigate to the **Top 10** sheet and add the **Department** filter and the newly added **Action (Category)** filter to the context using the drop-down menu on the fields. Once added to the context, these fields will be gray on the filters shelf. Now, you will see the top 10 items within the context of the selected **Department** and **Category**.

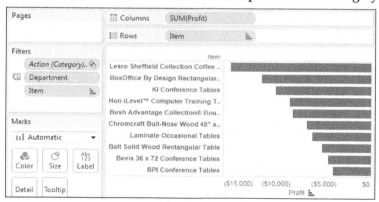

If you edit the action on the dashboard, the filter might be automatically updated and you may have to re-add it to the context.

3. While viewing the dashboard, navigate to **Dashboard | Actions...** from the menu. Add another filter action.

4. Name this action `Filter by Item` and set **Source Sheets** as **Top 10 Least Profitable Items** for the selected category and **Target Sheets** as both **When?** and **Where?**.

5. Set the action to run on **Select** and to exclude all values with the selection cleared. Click on **OK** on both the **Add Filter Action** and **Actions** dialog screens.

Go ahead and step through the actions by selecting a couple of different categories and a couple of different items. When you clear your selections, the target views on the dashboard will have all values excluded and will be blank.

This approach to building a dashboard, which starts with a single area of focus and then gradually expands the focus based on user interaction, has several advantages. First, with only one initial view, the dashboard loads faster and gives the audience a more immediate response. Second, the dashboard doesn't overwhelm the audience and guides them gradually through the data. This isn't the only approach to dashboard design and we'll see another example that is very different.

Observe how the final dashboard meets your objectives by telling a story:

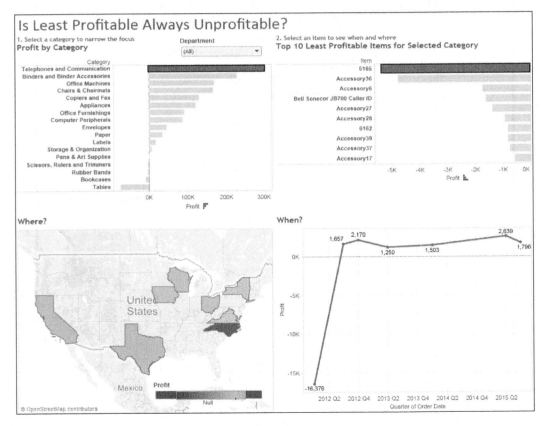

The user has selected **Telephones and Communications** as the category and then selected item **5165**, which is the least profitable item within the category. Should the management remove item **5165** from the inventory? Not without first considering that the item only lost profit in one instance. Granted, it was a large loss, but it was also a long time ago and every subsequent sale of the item resulted in gain.

When you look at the workbook of this chapter, you'll only see a tab at the bottom of the dashboard. The individual views have been hidden. Hiding tabs for sheets used in dashboards or stories is a great way to keep your workbook clean and guide your audience away from looking at sheets meant to be seen in the context of a dashboard or story. To hide a sheet, right-click on the tab and select **Hide Sheet**. To unhide a sheet, navigate to the dashboard or story using the sheet, right-click the sheet in the left-hand side pane, and uncheck **Hide Sheet**.

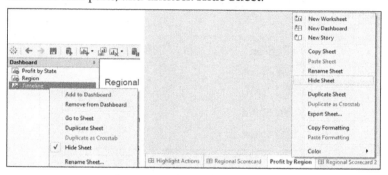

How actions work

You saw a few examples of actions used in dashboards. We'll now consider some details of how actions work in Tableau.

Tableau supports three kinds of actions:

- Filter actions
- Highlight actions
- URL actions

Certain actions are automatically generated by Tableau based on shortcuts. For example, you can select **Use as Filter** from the drop-down menu of a view on a dashboard, resulting in an automatically generated filter action. Enabling highlighting using the button on a discrete color legend or on the toolbar will automatically generate a highlight action.

You can create or edit dashboard actions by navigating to **Dashboard | Actions** from the menu. Let's consider the details of each type of action.

Filter actions

Filter actions are defined by a source sheet(s) that passes one or more dimensional values as filters to target sheets upon an action. Remember that every mark on a sheet is defined by the unique intersection of dimensional values. When an action occurs involving one or more of these marks, the dimensional values that comprise the mark(s) can be passed as filters to one or more target sheets.

When you create or edit a filter action, you will see options similar to these:

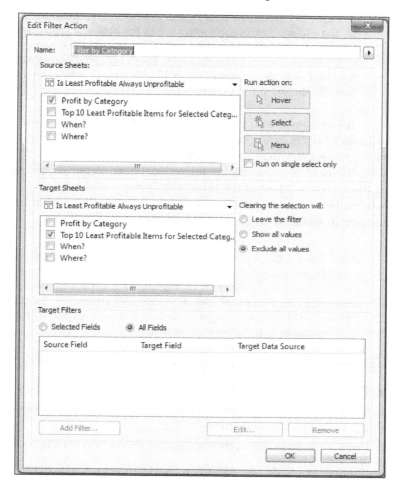

This screen allows you to:

- Name the filter
- Choose source and target sheets
- Set the action that triggers the filter and whether the selection of multiple marks, or only a single mark, initiates the action
- Choose what happens when the selection is cleared
- Specify which dimensions are used to pass filter values to the target sheet(s)

Use names for actions that will help you differentiate between multiple actions in the dashboard. Additionally, the name will be shown as a link in the tooltip when the action is set to be triggered via **Menu**. Use the arrow to the right of the name to insert special field placeholders. These will be dynamically updated with the values of the fields for a mark when the user sees the menu option in a tooltip.

You may select as many source and target sheets as you desire. However, if you specify precise **Target Filters** in the bottom section, the fields you select must be present in the target sheet (for example, on **Rows**, **Columns**, and **Detail**). You will receive a warning if a field is not available for one or more target sheets and the action will not fire for these sheets. Most of the time, your source and target will be the same dashboard. Optionally, you can specify a different target sheet or dashboard, which will cause the action to navigate to the target in addition to filtering.

When filter actions are defined at a worksheet level (when viewing a worksheet, navigate to **Worksheet | Actions** from the menu), then a menu item for that action will appear for every mark on every sheet that uses the same data source. You can use this to quickly create navigation between worksheets and from dashboards to individual worksheets.

Filter actions can be set to occur on any one of three possible actions:

- **Hover**: The user moves the mouse cursor over a mark
- **Select**: The user clicks on a mark or lassos multiple marks by clicking and dragging a rectangle around them
- **Menu**: The user selects the menu option for the action on the tooltip

Consider the following example of a filter action triggered when a bar is selected in the **Source** window:

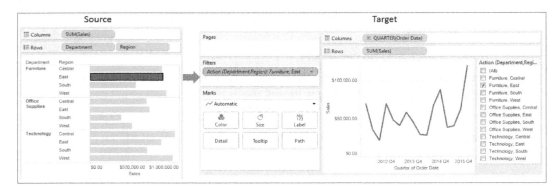

Each bar in the source is defined by the intersection of the **Department** and **Region** dimensions. When the bar for **Furniture/East** is selected, a single filter is set on the target. The filter is a set that contains the combination of dimensional values. Note that when the action filter is shown as a quick filter, it contains all possible combinations of **Department** and **Region** with **Furniture/East** checked.

By default, all dimensions present in the source view are used in a filter action. Optionally, you can specify which fields should be used. You can use the **Selected Fields** option to accomplish the following:

- Filters based on fewer dimensions. For example, if you only selected the **Region** field, then selecting the bar shown earlier would only pass the **East** region as a filter to the target (which would then still show data for all departments).

- Filter a target view using a different data source. You can even map the field to a field with a different name in the target view. For example, if the target used a data source where **East** was a possible value for a field named **Area**, you could map **Region** from the source to **Area** in the target.

A quick filter can be applied to multiple worksheets in a dashboard but cannot be applied to worksheets using different primary data sources. You can filer across data sources in a couple of different ways: action filters or parameters combined with calculated fields as filters in each data source. Action filters have the advantage of allowing multiselect and are based on views that dynamically change with the data. Parameters are a single selection only and have a static set of selection options. However, parameter controls more closely resemble quick filter controls and have greater flexibility for use in views.

Highlight actions

Highlight actions do not filter target sheets. Instead, they cause marks that are defined, at least in part, by the selected dimensional value(s) to be highlighted in the target sheets. The options for highlight actions are very similar to filter actions, with the same options for source and target sheets and which event triggers the action.

Consider a dashboard with the following two views and a highlight action based on the **Region** field. When the action is triggered for the **Central** region, all marks defined by **Central** (the map has **Region** on the level of detail) are highlighted:

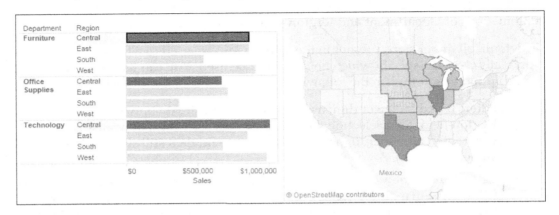

URL actions

URL actions allow you to dynamically generate a URL based on an action and open it within a web object in the dashboard or in a new browser window or a tab. URL actions can be triggered by the same events as filter and highlight actions. The name of the URL action differentiates it and will appear as the link when used as a menu.

The URL includes any hardcoded values you enter as well as placeholders accessible via the arrow to the right of the URL textbox. These placeholders include fields and parameters. The values will be dynamically inserted into the URL string when the action is triggered based on the values for the fields that make up the selected mark(s) and current values for parameters.

If you have included a web object in the dashboard, the URL action will automatically use it as the target. Otherwise, the action opens a new browser window (when the dashboard is viewed in Desktop or Reader) or a new tab (when the dashboard is viewed in a web browser).

Some web pages display a different behavior when viewed in **iframes**. The browser object does not use iframes in Tableau Desktop or Tableau Reader but uses them when the dashboard is published to Tableau Server, Tableau Online, or Tableau Public. You will want to test URL actions based on how your dashboards will be viewed by your audience.

Example – a regional scorecard

We'll consider another example dashboard that demonstrates slightly different objectives. Let's say everyone in the organization has agreed upon a key performance indicator of the profit ratio. Furthermore, everyone agrees that the cutoff between good and poor profit ratio is 15 percent but would like to have the option of adjusting the value dynamically to see whether other targets would be better.

Consider the following dashboard:

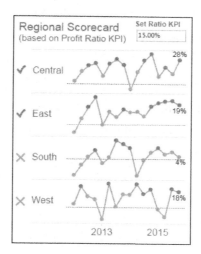

This dashboard allows your audience to very quickly evaluate the performance of each region. Additionally, it is easy to see the historical performance evaluated by the standard of the KPI. The dashboard provides some interactivity with the KPI parameter and tooltips. Additional drilldown into other dashboards or views could be provided if desired. If this view were published on Tableau Server, it is not unreasonable to think that regional managers might subscribe to the view and receive a daily e-mail containing an up-to-date image of this dashboard.

One of the first things you might notice is that this is a very small dashboard and seemingly very simple. Depending on your objectives, audience, and mode of delivery, you might decide upon a certain size of dashboard. For example, the preceding dashboard might work well on a mobile device. You can specify any number of predefined resolutions (laptop, desktop, iPad, and so on), specify a dynamically sized dashboard (**Automatic** or **Range**), or specify exact dimensions (**Exactly**).

Dynamically sized dashboards have the advantage of filling up what might otherwise be empty space. However, fixed-size dashboards have the advantage of giving you more precise control over layout, and you avoid some of the unfortunate surprises that can occur when a dynamically sized dashboard doesn't look exactly the way you designed it.

The preceding dashboard consists of two views, a parameter control, and a text object. To create a similar dashboard, follow these steps:

1. Create a float type parameter named Profit Ratio Cutoff set to an initial .15 value formatted as a percent.

2. Create a calculation named Profit Ratio with the SUM([Profit]) / SUM([Sales]) code. This is an aggregate calculation that will divide the sum of profit by the sum of sales at the level of detail defined in the view.

3. Create a second calculation named Profit Ratio KPI with the code:

```
IF
    [Profit Ratio] >= [Profit Ratio Cutoff]
THEN "Good"
ELSE "Poor"
END
```

This code will compare the profit ratio to the parameterized cut-off value. Anything equal to or above the cutoff will get the value of Good and everything below will get the value of Poor.

4. Create a new sheet named Regions Overall. The view consists of **Region** on **Rows** (with **Show Header** unchecked via the drop-down menu) and the **Profit Ratio KPI** field on both **Shape** and **Color**. You'll observe that the shapes have been edited to use checkmarks and Xes and the color palette uses color-blind-safe blue and orange. The worksheet formatting has been customized to remove row and column dividers.

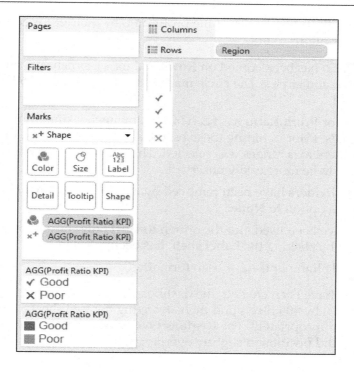

Create another view named Regions Sparklines.

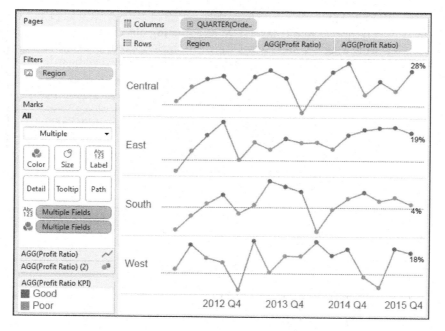

This view is basically a time series of the profit ratio by quarter for each region. There are a few customizations to note:

- **Profit Ratio** has been copied on **Rows** and uses a synchronized dual axis to plot a line and circles. The circle marks have been color-encoded by **Profit Ratio KPI**.

- The axis for **Profit Ratio** has been hidden by unchecking **Show Header** for these fields. Prior to hiding the axes, the axis was edited and the option for independent axis ranges was selected. This allows the overall shape of each sparkline to be more easily observed.

- Column dividers have been removed by navigating to **Format | Borders** and setting the option to **None**.

- Labels have been used and the option to label only the end of the line, available by clicking the **Label** shelf, has been selected.

- Fonts and alignments have been formatted.

Once both views have been created, the dashboard can be constructed by specifying an exact size of 300 by 400 pixels and then arranging the two sheets, parameter control, and text appropriately. The **Regions Overall** sheet has been added as a floating object and positioned slightly overlapping to the left of the **Regions Sparklines** view.

By default, all objects added to the dashboard are tiled. Tiled objects appear beneath floating objects. Any object can be added to the dashboard as a floating object by switching the toggle under **New Objects** in the left window or by holding *Shift* while dragging the objects to the dashboard.

Existing objects can be switched between floating and tiled by holding *Shift* while moving the object or using the drop-down caret menu. The drop-down caret menu also gives options to adjust the floating order of objects. Additionally, floating objects can be resized and positioned with pixel precision by selecting the floating object and using the positioning and sizing controls in the lower-left section.

Story points

The story points feature allows you to tell a story using interactive snapshots of dashboards and views. The snapshots become points in a story. This allows you to construct a guided narrative or even an entire presentation.

Let's consider an example in which story points might be useful. Executive managers are pleased with the regional scorecard dashboard you developed. Now, they want you to make a presentation to the board and highlight some specific issues for the **East** region. With minimal effort, you can transform even a simple scorecard into an entire story, following steps such as these:

1. First, we'll extend the scorecard with one additional view. Create a view similar to this one by double-clicking on the **State** field, adding **Profit Ratio KPI** to color, and including **Profit Ratio** on the **Detail** level of the **Marks** card. If needed, edit the color legend to use a **Blue / Orange** palette. If desired, increase the washout of the map to **100%** (from the menu, navigate to **Map | Options**) to remove the background clutter.

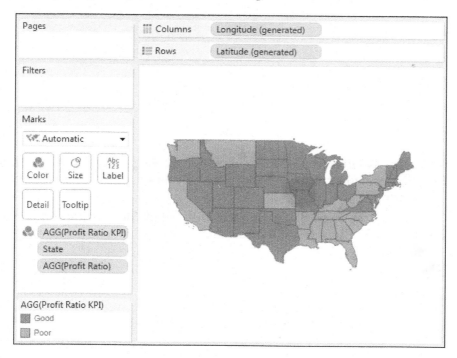

2. Duplicate the **Regional Scorecard** dashboard and include the new view. You may need to increase the width to **600** pixels. If needed, remove the color legend.

3. Use the dropdown on the **Region** sparklines sheet to add a quick filter for **Region**. Use the dropdown on the **Region** quick filter and navigate to **Apply to Worksheets | Selected Worksheets....** and then choose **All on Dashboard**. Rearrange the dashboard as needed until it looks similar to this:

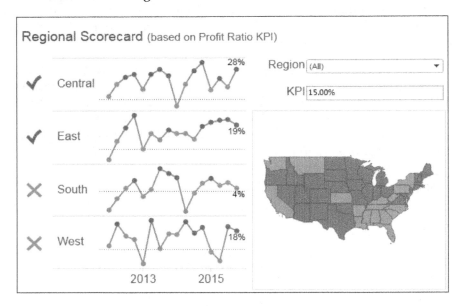

4. Select **Use as Filter** from the dropdown on the **Region** sparklines view to use that view as a filter in the dashboard. At this point, you are ready to build your story.

5. Create a new story by navigating to **Story | New Story** from the menu or using the **New Story** tab at the bottom.

The **Story** interface consists of a sidebar with all visible dashboards and views. At the top, you'll see **Story Title**, which can be edited. Each new point in the story will appear as a navigation box with text that can also be edited. Clicking on the box will give you access to the story point, to which you can add a single dashboard or view:

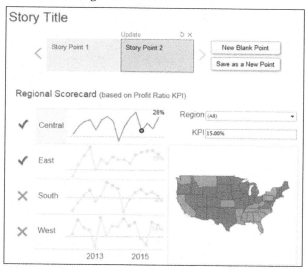

You can create new story points using the **New Blank Point** button (for a new story point), the **Duplicate** button (which will create a duplicate snapshot of the currently selected story point), or the **Save as a New Point** button (which will capture the current state of the dashboard as a new story point).

Clicking on a story point navigation box will bring up the snapshot of the view or dashboard for that story point. You may interact with the dashboard by, for example, making selections, changing quick filters, changing parameter values, and adding annotations. Changing any aspect of the dashboard will present you with an option to update the existing story point to the current state of the dashboard. Alternately, you can use the revert button above the navigation box to return to the original state of the dashboard. Clicking on the **X** button will remove the story point.

Each story point contains an entirely independent snapshot of a dashboard. Quick filter selections, parameter values, selections, and annotations will be remembered for a particular story point but will have no impact on other story points or any other sheet in the dashboard.

You may rearrange story points by dragging and dropping the navigation boxes.

We'll build the story by completing the following steps:

1. Give the story the title East Region Analysis.

2. Add the dashboard you created earlier to the first story point. We'll give the story point the The East region is good overall text.

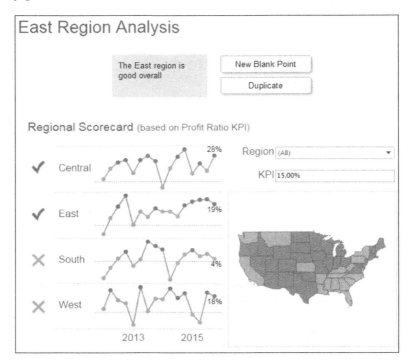

3. Click on the **Duplicate** button to copy the current story point. Give the second story point the The East region has seen ups and downs text and select only **East** in the **Region** quick filter. Note that switching between the two story points reveals that quick filters are not shared between story points.

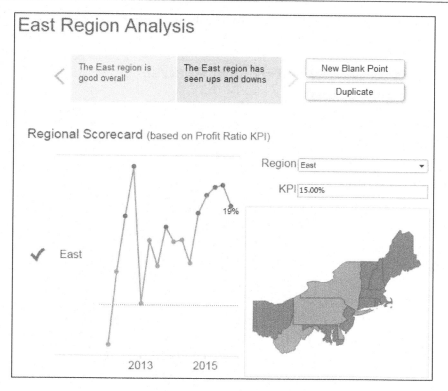

4. Click on the **Duplicate** button to copy the current story point again. Give the second story point the `Only one quarter would have met a goal of 25%` text and set the KPI to `.25`.

5. Add a mark annotation to the single quarter that is above 25 percent. Observe that the KPI and the annotation are specific to this story point and are not shared with any other story points.

In **Presentation Mode**, the buttons to add, duplicate, update, or remove story points are not shown. Your final story should look similar to this:

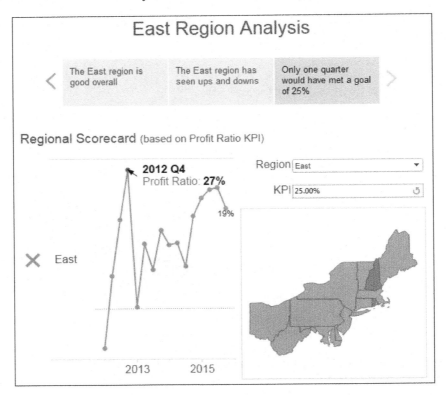

Take some time to walk through the presentation. Clicking on navigation boxes will show that story point. You can fully interact with the dashboard in each story point. This way, you can answer questions on the fly, dig into details, and then continue through the story.

A great way to learn dashboard techniques (and data visualization techniques in general) is to subscribe to *Viz of the Day* (http://www.tableau.com/public/community/viz-of-the-day). A new visualization, dashboard, or story is selected and highlighted each day. You can download the workbook and explore the various techniques some of the best designers have used.

Summary

When you are ready to share your discovery and analysis, you will most likely use dashboards to relate the story to your audience. The way in which you tell the story will depend on your objectives as well as your audience and the mode of delivery. Using a combination of views, objects, parameters, quick filters, and legends, you can create an incredible framework to tell a data story. By introducing actions and interactivity, you can invite your audience to participate. Story points will allow you to bring together many snapshots of dashboards and views to craft and present entire narratives.

Next, we'll turn our attention to some deeper analysis with trends, distributions, and forecasting.

Summary

8

Adding Value to Analysis – Trends, Distributions, and Forecasting

Sometimes, quick data visualization needs a little deeper analysis. For example, a simple scatterplot can reveal outliers and clusters of values. However, often, you want to understand the distribution. A simple time series helps you see the rise and fall of a measure over time. But many times, you want to see the trend or make predictions of future values.

Tableau enables you to quickly enhance your data visualizations with statistical analysis. Built-in features, such as trend models, distributions, and forecasting, allow you to quickly add value to your visual analysis. Additionally, Tableau integrates with R, an extensive statistical platform that opens up endless options for the statistical analysis of your data. This chapter will cover built-in statistical models and analysis.

This chapter will cover the following topics:

- Trending
- Forecasting
- Distributions

We'll take a look at these concepts in the context of a few examples using some sample datasets. You can follow and reproduce these examples using this chapter's workbook.

Trends

Let's say you are analyzing populations of various countries using the World Population dataset in the provided workbook. This dataset produces one record containing the population for each country for each year from 1960 to 2013. Create a view similar to the one shown in the following screenshot, which shows you the change in population over time for Afghanistan and Australia. You'll notice that **Country Name** has been filtered and added to the **Color** and **Label** shelves.

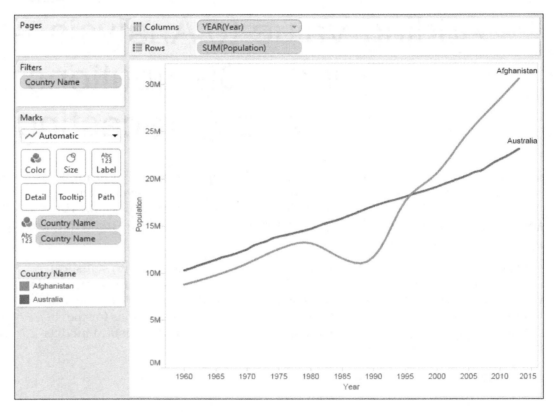

From this visualization alone, you can make several interesting observations. The growth of the two countries' populations was fairly similar up to 1980. At that point, the population of Afghanistan went into decline until 1988 when the population of Afghanistan started to recover. At some point around 1996, the population of Afghanistan exceeded that of Australia. The gap has grown wider ever since.

While we have a sense of the two trends, they become even more obvious when we see them. Tableau offers several ways to add trend lines:

- From the menu, navigate to **Analysis | Trend Lines | Show Trend Lines**
- Right-click on an empty area in the pane of the view and select **Show Trend Lines**
- Switch to the **Analytics** tab in the left-hand side pane and drag and drop **Trend Line** on the trend model of your choice (we'll use **Linear** for now and discuss the others later in this chapter)

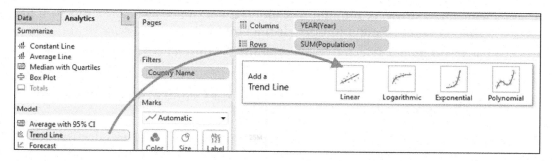

Once you have added the trend line, your view should look like this:

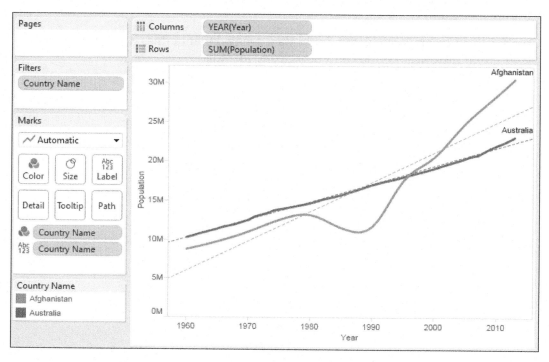

Trends are calculated by Tableau after querying the data source. Trend lines are drawn based on various elements in the view:

- **The two fields that define x and y coordinates**: The last (right-most) field on **Rows** and **Columns** will define the axes that give Tableau x and y coordinates to calculate various trend models. In order to show trend lines, you must use a continuous (green) field or discrete (blue) date fields and have one such field on both **Rows** and **Columns**. If you use a discrete (blue) date field to define headers, the other field must be continuous (green).

- **Additional fields that create multiple, distinct trend lines**: Discrete (blue) fields on the **Rows**, **Columns**, or **Color** shelves can be used as factors to split a single trend line into multiple, distinct trend lines.

- **The trend model selected**: We'll examine the differences in models in the next section.

Observe in the view that there are two trend lines. As **Country Name** is a discrete (blue) field on **Color**, it defines a trend line per color by default.

Earlier, we observed that the population of Afghanistan increased and decreased within various historical periods. Notice that the trend lines are calculated along the entire date range. What if we want to see different trend lines for these time periods? We can force Tableau to draw distinct trend lines using a discrete field on **Rows**, **Columns**, or **Color**.

Go ahead and create a calculated field called `Period` that defines discrete values for the different historical periods and using code like this:

```
IF Year([Year]) <= 1979
  THEN "1960 to 1979"
ELSEIF Year([Year]) <= 1988
  THEN "1980 to 1988"
ELSE "1988 to 2013"
END
```

When you place **Period** on columns, you'll get a header for each time period, which breaks the lines and causes separate trends to be shown for each time period. You'll also observe that Tableau keeps the full date range in the axis for each period. You can set an independent range by right-clicking on one of the date axes, selecting **Edit Axis**, and then checking the option for **Independent axis range for each row or column**.

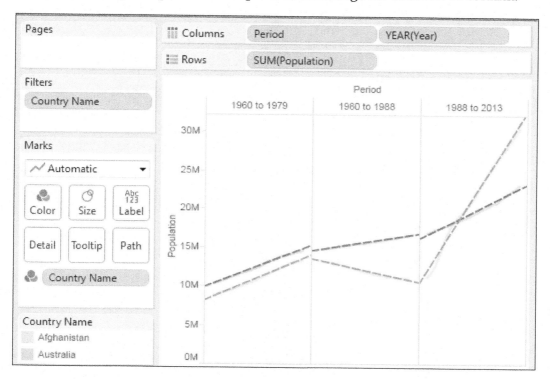

In this view, transparency has been applied to **Color** and the trend lines have been formatted to make them stand out. Additionally, the axis for **Year** was hidden (by unchecking the **Show Header** option on the field). Now, you can clearly see the difference in trends for different periods of time. Australia's trends only change slightly in each period. Afghanistan's trends were quite different.

Customizing trend lines

Let's take a look at another example that will allow us to consider various options for trend lines. Create a new sheet and use the `Real Estate` data source connection to create a view similar to this one:

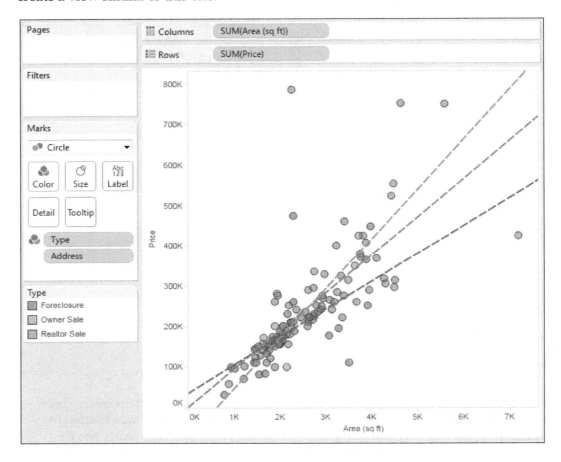

Here, we created a scatterplot with the sum of **Area** on **Columns** to define the *x* axis and the sum of **Price** on **Rows** to define the *y* axis. **Address** has been added to the level of detail on the **Marks** card to define the level of aggregation. So, each mark on the scatterplot is a distinct address at a location defined by the area and price. **Type** has been added to **Color**. We've also shown the trend lines and are getting one trend line per color by default. Assuming a good model, the trend lines demonstrate how much and how quickly **Price** is expected to rise with an increase in **Area**.

In this dataset, we have two fields, **Address** and **ID**, either of which define a unique record. Adding one of these fields to the level of detail effectively disaggregates the data and allows us to plot a mark for each address. Sometimes, you may not have a field in the data that defines uniqueness. In these cases, you can disaggregate the data by unchecking **Aggregate Measures** from the **Analysis** menu.

Alternately, you can use the drop-down menu on each of the measure fields on **Rows** and **Columns** to change them from measures to dimensions while keeping them continuous. As dimensions, each individual value will define a mark. Keeping them continuous will retain the axes required for trend lines.

Let's consider some of the options available for trend lines. You can edit trend lines by using the menu and navigating to **Analysis | Trend Lines | Edit Trend Lines** or clicking/right-clicking on a trend line and then selecting **Edit**. When you do this, you'll see a dialog box similar to this:

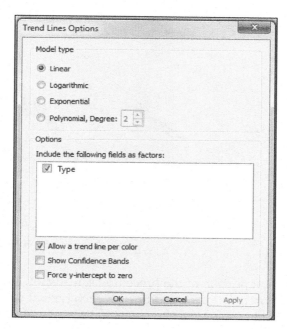

Here, you have options to select a model type, including applicable fields as factors in the model, allowing discrete colors to define distinct trend lines, showing confidence bands, and forcing the *y* intercept to zero. Experiment with the options for a bit. Notice how either removing the **Type** field as a factor or unchecking the **Allow a trend line per color** option results in a single trend line.

You can also see the result of excluding a field as a factor in the following view, where **Type** has been added to **Rows**:

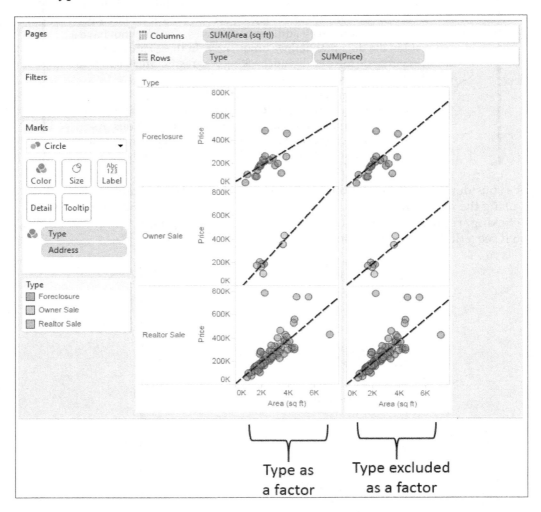

As represented in the left portion of the preceding screenshot, **Type** is included as a factor. This results in a distinct trend line for each type of sale. When **Type** is excluded as a factor of the same trend line, which is the overall trend for all types, a trend line is drawn three times. This technique can be quite useful to compare subsets of data with the overall trend.

Trend models

We'll return to the original view and stick with a single trend line as we consider the trend models available. The following models can be selected from the **Trend Line Options** window:

- **Linear**: We'll use this model if we assume that as **Area** increases, **Price** will increase at a constant rate. No matter how high **Area** increases, we'll expect **Price** to increase such that new data points fall close to the straight line.

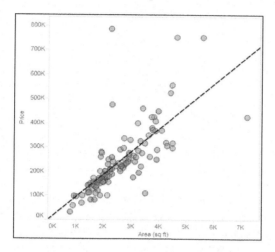

- **Logarithmic**: We'll use this model if we believe that there is a "law of diminishing returns" in effect. That is, area can only increase to a certain extent before buyers stop paying much more:

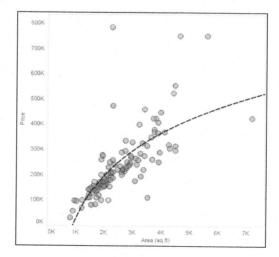

- **Exponential**: We'll use this model to test the idea that each additional increase in area results in a dramatic (exponential) increase in price:

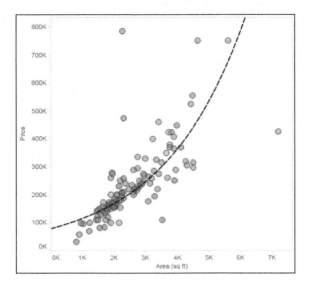

- **Polynomial**: We'll use this model if we feel the relationship between **Area** and **Price** is complex and follows more of an S-shaped curve, where, though initially increasing the area dramatically increases the price, at some point the price levels. You can set the degree of the polynomial model anywhere from 2 to 8. The trend line shown here is a third-degree polynomial:

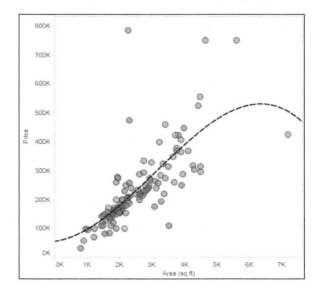

Analyzing trend models

It can be useful to observe trend lines, but often, we'll want to understand whether the trend model we've selected is statistically meaningful. Fortunately, Tableau gives us some visibility into trend models and calculations.

Simply hovering over a single trend line will reveal the calculation as well as **P-value** for that trend line.

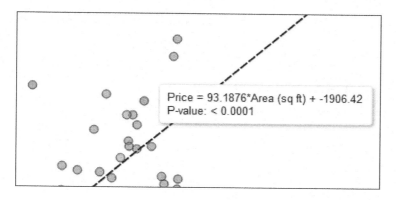

A p-value is a statistical concept that describes the probability that the results of assuming no relationship between values (random chance) are at least as close as results predicted by the trend model. A p-value of 5 percent (.05) will indicate a 5 percent random chance describing the relationship between values as well as the trend model. This is why p-values of 5 percent or less are considered to indicate a significant trend model. If your p-value is higher than 5 percent, then you should not consider that trend to significantly describe any correlation.

Additionally, you can see a much more detailed description of the trend model by navigating to **Analysis | Trend Lines | Describe Trend Model…** from the menu or using the similar menu from a right-click on the view's pane. When you view the trend model, you will see the **Describe Trend Model** window:

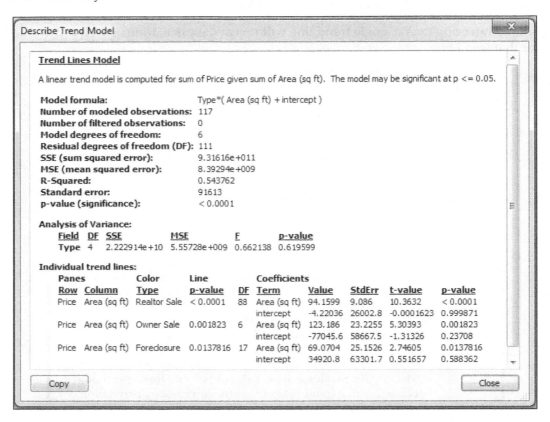

Describe Trend Model

Trend Lines Model

A linear trend model is computed for sum of Price given sum of Area (sq ft). The model may be significant at p <= 0.05.

Model formula:	Type*(Area (sq ft) + intercept)
Number of modeled observations:	117
Number of filtered observations:	0
Model degrees of freedom:	6
Residual degrees of freedom (DF):	111
SSE (sum squared error):	9.31616e+011
MSE (mean squared error):	8.39294e+009
R-Squared:	0.543762
Standard error:	91613
p-value (significance):	< 0.0001

Analysis of Variance:

Field	DF	SSE	MSE	F	p-value
Type	4	2.22914e+10	5.55728e+009	0.662138	0.619599

Individual trend lines:

Panes			Line		Coefficients				
Row	Column	Color Type	p-value	DF	Term	Value	StdErr	t-value	p-value
Price	Area (sq ft)	Realtor Sale	< 0.0001	88	Area (sq ft)	94.1599	9.086	10.3632	< 0.0001
					intercept	-4.22036	26002.8	-0.0001623	0.999871
Price	Area (sq ft)	Owner Sale	0.001823	6	Area (sq ft)	123.186	23.2255	5.30393	0.001823
					intercept	-77045.6	58667.5	-1.31326	0.23708
Price	Area (sq ft)	Foreclosure	0.0137816	17	Area (sq ft)	69.0704	25.1526	2.74605	0.0137816
					intercept	34920.8	63301.7	0.551657	0.588362

Copy   Close

You can also get a trend model description in the worksheet description, which is available from the **Worksheet** menu or by pressing *Ctrl + E*. The worksheet description includes quite a bit of other useful summary information about the current view.

The wealth of statistical information shown in the window includes a description of the trend model, the formula, the number of observations, and the p-value for the model as a whole as well as for each trend line. Note that in the window shown in the preceding screenshot, the **Type** field was included as a factor that defined three trend lines. At times, you may observe that the model as a whole is statistically significant even though one or more trend lines may not be.

Additional summary statistical information can be displayed in Tableau Desktop for a given view by showing the summary. From the menu, navigate to **Worksheet | Show Summary**. The information displayed in the summary can be expanded using the drop-down menu on the **Summary** card.

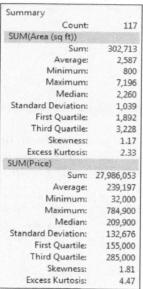

Summary	
Count:	117
SUM(Area (sq ft))	
Sum:	302,713
Average:	2,587
Minimum:	800
Maximum:	7,196
Median:	2,260
Standard Deviation:	1,039
First Quartile:	1,892
Third Quartile:	3,228
Skewness:	1.17
Excess Kurtosis:	2.33
SUM(Price)	
Sum:	27,986,053
Average:	239,197
Minimum:	32,000
Maximum:	784,900
Median:	209,900
Standard Deviation:	132,676
First Quartile:	155,000
Third Quartile:	285,000
Skewness:	1.81
Excess Kurtosis:	4.47

Tableau also gives you the ability to export data, including data related to trend models. This allows you to more deeply, and even visually, analyze the trend model itself. Let's analyze the third-degree polynomial trend line of the real estate price and area scatterplot without any factors. To export data related to the current view, use the menu and navigate to **Worksheet | Export | Data**. The data will be exported as a Microsoft Access Database (.mdb) and you will be prompted as to where to save the file.

On the **Export Data to Access** screen, specify an access table name and select whether you wish to export data from the entire view or the current selection. You may also specify that Tableau should connect to the data. This will generate the data connection and make it available with the specified name in the current workbook.

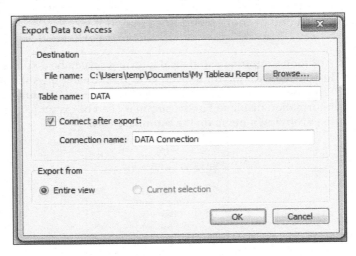

The new data source connection will contain all the fields that were present in the original view as well as additional fields related to the trend model. This allows us to build a view such as the following using residuals and predictions:

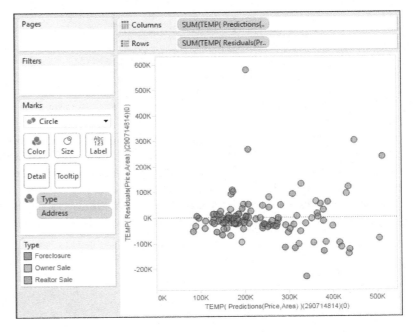

A scatterplot of predictions and residuals allows you visually see how far each mark was from the location predicted by the trend line. It also allows you to see whether residuals are distributed evenly on either side of a zero. An uneven distribution would indicate problems with the trend model.

You can include this new view along with the original one in a dashboard to explore the trend model visually. Use the highlight button on the toolbar to highlight the **Address** field.

With the highlight action defined, selecting marks in one view will allow you to see them in the other. You could extend this technique to export multiple trend models and dashboards to evaluate several trend models at the same time, as shown in the following screenshot:

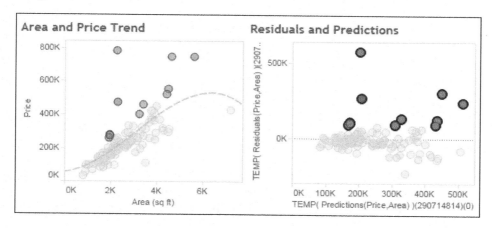

 You can achieve even more sophisticated statistical analysis, leveraging Tableau's ability to integrate with R. R is an open source statistical analysis platform and a programming language with which you can define advanced statistical models. R functions can be called from Tableau using special table calculations (all of which start with `SCRIPT_`). These functions allow you to pass expressions and values to a running R Server that will evaluate the expressions using built-in libraries or custom-written R scripts and return results to Tableau.

You can learn more about Tableau and R integration from this whitepaper (you will need to register a free account first): http://www.tableausoftware.com/learn/whitepapers/using-r-and-tableau

Distributions

Analyzing distributions can be quite useful. We've already seen that certain table calculations are available to determine statistical information such as averages, percentiles, and standard deviations. Tableau also makes it easy to quickly visualize various distributions including confidence intervals, percentages, percentiles, quantiles, and standard deviations.

You may add any of these visual analytic features using the **Analytics** tab (alternately, you can right-click on an axis and select **Add Reference Line**). Just like reference lines and bands, distribution analytics can be applied within the scope of a table, pane, or cell. When you drag and drop the desired visual analytic, you'll have options to select the scope and the axis. In the following example, we've dragged and dropped **Distribution Band** from the **Analytics** tab onto the scope of **Pane** for the axis defined by **Sum(Price)**:

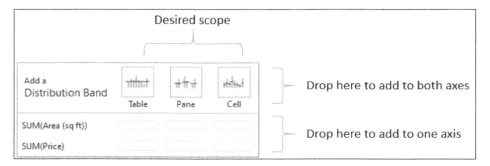

Once you have selected the scope and axis, you will be given options to change settings. You may also edit lines, bands, distributions, and box plots by right-clicking on the analytic feature in the view or by right-clicking on the axis. Here, we'll define settings for one and two standard deviations above and below the mean:

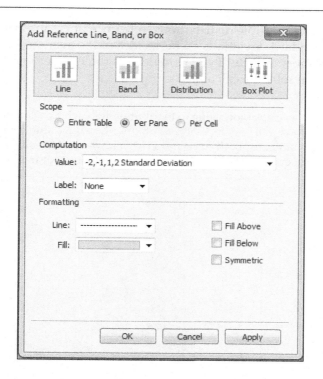

Each specific **Distribution** option specified in the **Value** dropdown under **Computation** has unique settings. **Confidence Interval**, for example, allows you to specify a percent value for the interval. **Standard Deviation** allows you to enter a comma-delimited list of values that describe how many standard deviations are used, and at what intervals. This, for example, is the result of specifying standard deviations of **-2, -1, 1, 2**:

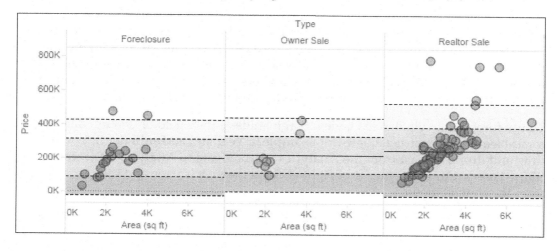

Each axis can support multiple distributions, reference lines, and bands. Here, first and second standard deviations on both sides of the average (the solid line) are shown. You'll notice that the **Type** field defines three panes and the standard deviations have been set to be calculated per pane.

On a scatterplot, using a distribution for each axis can yield a very useful way to analyze outliers. Showing a single standard deviation for both **Area** and **Price** allows you to easily see properties that fall within norms for both, one, or neither.

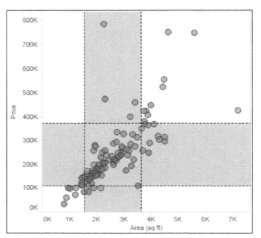

Forecasting

As we've seen, trend models make predictions. Given a good model, you expect additional data to follow the trend. When the trend is over time, you can get an idea about where future values may fall. However, predicting future values often requires a different type of model. Factors such as seasonality can make a difference not predicted by a trend alone. Starting with version 8.0, Tableau includes built-in forecasting models that can be used to predict and visualize future values.

To use forecasting, you'll need a view that includes a date field or enough date parts for Tableau to reconstruct a date (for example, a **Year** and a **Month** field). You may drag and drop **Forecast** from the **Analytics** tab, navigate to **Analysis | Forecast | Show Forecast** from the menu, or right-click on the view's pane and select the option from the context menu.

Here, for example, is the view of the population growth of Afghanistan and Australia with forecasts shown over time:

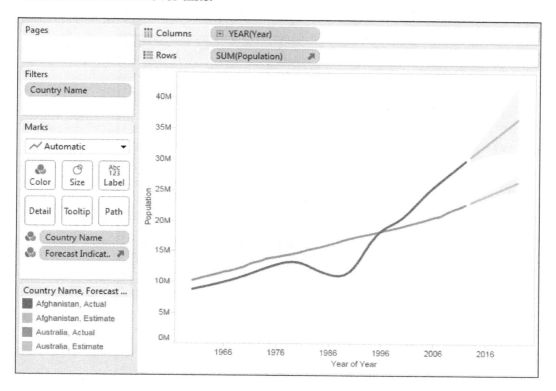

Note that when you show the forecast, Tableau adds a forecast icon to the **SUM(Population)** field on **Rows** to indicate that the measure is being forecast. Additionally, Tableau adds a new special **Forecast Indicator** field to **Color** so that forecast values are differentiated from actual values in the view.

 You can move the **Forecast Indicator** field or even copy it (hold *Ctrl* while dragging and dropping) to other shelves to further customize your view.

When you edit the forecast by navigating to **Analysis | Forecast | Forecast Options...** from the menu or when you use the right-click context menu on the view, you will be presented with various options to customize the trend model, like this:

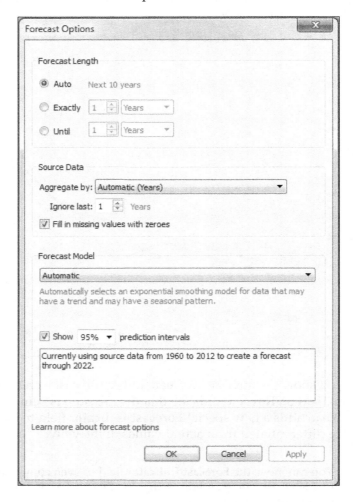

Here, you have options to set the length of the forecast, determine aggregations, customize the model, and set whether you wish to show prediction intervals. The forecast length is set to **Auto** by default, but you can extend the forecast by a custom value.

The options under **Source Data** allow you to optionally specify a different grain of data for the model. For example, your view might show a measure by year but you could allow Tableau to query the source data to retrieve values by month and use a finer grain to potentially achieve better results.

Tableau's ability to separately query the data source to obtain data at a finer grain for more precise results works well with relational data sources. However, OLAP data sources are not compatible with this approach, which is one reason forecasting is not available when working with cubes.

By default, the last value is excluded from the model. This is useful when you are working with data where the most recent time period is incomplete. For example, when records are added daily, the last (current) month is not complete until the final records are added on the last day of the month. Prior to this last day, the incomplete time period might skew the model unless it is ignored.

The model itself can be set to **Automatic** with or without seasonality or can be customized to set options for seasonality and trend. To understand the options, consider the following view of **Sales** by **MONTH** from the Superstore sample data:

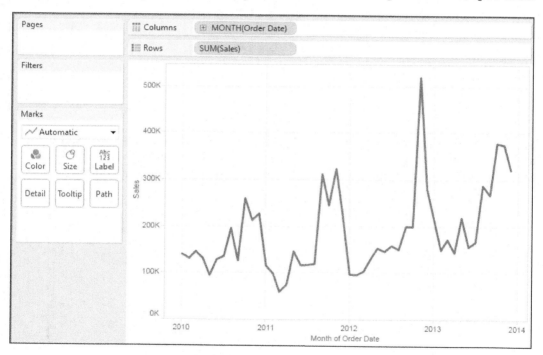

The data displays a distinct cyclical or seasonal pattern. This is very typical for retail sales. The following are the results of selecting various custom options:

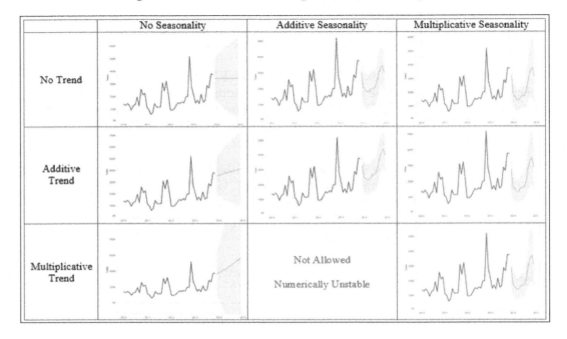

Much like trends, forecast models and summary information can be accessed using the menu. Navigating to **Analysis | Forecast | Describe Forecast** will display a window with tabs for both the summary and details concerning the model.

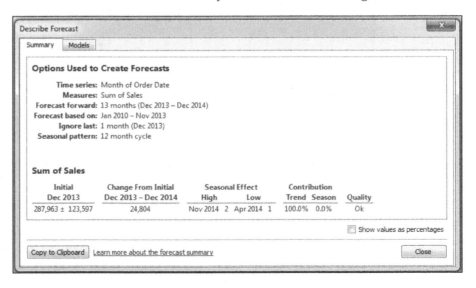

Clicking on the link at the bottom of the window will give you much more information on the forecast models used in Tableau.

 Forecast models are only enabled with a certain set of conditions. If the option is disabled, ensure that you are connected to a relational database and not OLAP, are not using table calculations, and have at least five data points.

Summary

Tableau provides an extensive set of features to add value to your analysis. Trend lines allow you to more precisely identify outliers, determine which values fall within the predictions of certain models, and even make predictions of where measurements are expected. Tableau gives you extensive visibility into the trend models and even allows you to export data containing trend model predictions and residuals. Distributions are useful to understand how measurements are distributed. Forecasting allows a complex model of trends and seasonality to predict future results. Having a good understanding of these tools will give you the ability to clarify and validate your initial visual analyses.

Next, we'll turn our attention back to the data. We considered very early on how to connect to data, and we've been working with data ever since. However, we've spent most of our time working with clean, well-structured data. In the next chapter, we'll consider how to deal with messy data.

9
Making Data Work for You

Up to this point, most of the examples we looked at in this book assume that data is structured well and is fairly clean. Data in the real world isn't so pretty at times. Maybe it's messy or it doesn't have a good structure. Maybe it has missing values or duplicate values. Or maybe it is at the wrong level of detail.

How can you deal with this messy data? Tableau offers quite a bit of flexibility to address data issues within the tool. We'll take a look at some of the features and techniques that will enable you to overcome data structure obstacles. Having a good understanding of what data structures work well with Tableau is a key factor to understand how you will be able to resolve certain issues.

In this chapter, we'll focus on some principles that structure data to work well with Tableau as well as some specific examples of how to address common data issues. This chapter will cover the following topics:

- Structuring data for Tableau
- Techniques to deal with data structure issues
- An overview of advanced fixes for data problems

Structuring data for Tableau

We've already seen that Tableau can connect to nearly any data source. Whether it's a built-in direct connection or ODBC or whether it's using the Tableau Data Extract API to generate an extract, no data is off-limits. However, there are certain structures that make data easier to work with in Tableau.

There are two key ways to ensure a good data structure that works well with Tableau:

- Every record of a source data connection should be at a meaningful level of detail
- Every measure contained in the source should match the level of detail or possibly be at a higher level of detail; however, it should never be at a lower level of detail

For example, you may have one record per class in a school. Within the record, you may have three measures: the average GPA of the class, the number of students in the class, and the average GPA of the school. The first two measures are at the same level of detail as the individual record of data (per student). The GPA for the school is at a higher level of detail (per school). As long as you are aware of this, you can perform a careful analysis. However, you would have a data structure issue if you tried to store each student's GPA in the class record.

Understanding the level of detail of the source (often referred to as **granularity**) is vital. Every time you connect to a data source, the very first question you should ask and answer is this: What does a single record represent? If, for example, you were to drag and drop the **Number of Records** field into the view and observed 4,000 records, then you should be able to complete this statement: "I have 4,000 _____." It could be 4,000 students or 4,000 test scores for 1,000 distinct students. Having a good grasp of the granularity of the data will help you avoid poor analysis and allow you to determine whether you even have the data necessary for your analysis.

Under the key principle of the granularity of the data, there are certain data structures that allow you to work seamlessly and efficiently in Tableau. Sometimes, it is preferable to restructure the data at the source using tools specifically designed for ETL (**which stands for Extract, Transform**, and **Load**) or reshaping data. Sometimes, restructuring the source data isn't possible or is not feasible. We'll take a look at some options in Tableau for these cases. For now, let's consider what kinds of data structure work well with Tableau.

Good structure – tall and narrow instead of short and wide

The two keys of good structure mentioned earlier should result in a data structure where a single measure is contained in a single column. You may have multiple different measures, but any single measure should not be divided into multiple columns. Often, the difference is described as tall data versus wide data. **Tall data** describes a structure in which each distinct measure in a row is contained in a single column. Tall data often results in more rows and fewer columns. **Wide data** describes a structure in which a measure in a single row is spread over multiple columns. Wide data often results in fewer rows with more columns.

Let's consider an example. Wide data looks like this table of population numbers:

Country name	1960	1961	1962	1963	1964
Afghanistan	8,774,440	8,953,544	9,141,783	9,339,507	9,547,131
Australia	10,276,477	10,483,000	10,742,000	10,950,000	11,167,000

Note that the table contains a row for every country. However, the measure (population) is not stored in a single column for each country. Instead, the measure is stored per country per year. This data is wide because it has a single measure (population) that is being divided into multiple columns (a column for each year). The wide data table violates the second key of good structure in that the measure is at a lower level of detail than the individual record.

Now, consider the following table, which represents the same data in a tall structure:

Country name	Year	Population
Afghanistan	1960	8,774,440
Afghanistan	1961	8,953,544
Afghanistan	1962	9,141,783
Afghanistan	1963	9,339,507
Afghanistan	1964	9,547,131
Australia	1960	10,276,477
Australia	1961	10,483,000
Australia	1962	10,742,000
Australia	1963	10,950,000
Australia	1964	11,167,000

Now, we have more rows (a row for each year for each country). Individual years are no longer separate columns and population measurements are no longer spread across these columns. Instead, one single column represents the year and another single column represents the population. The number of rows has increased while the number of columns has decreased. Now, the measure of population is at the same level of detail as the individual row.

You can easily see the difference between wide and narrow data in Tableau. Here is what the wide table of data looks like in the left-hand side **Data** window:

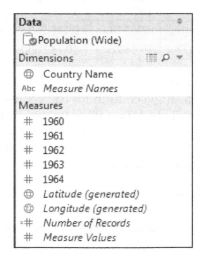

As we'd expect, Tableau treats each column in the table as a separate field. The wide structure of the data works against us. We end up with a separate measure for each year. If you wanted to plot a line graph of population per year, you might struggle. What dimension represents the date? What single measure can you use for population?

In contrast, tall data looks like this:

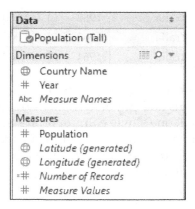

This data source is much easier to work with. If you want a line chart of population by year, you can simply drag and drop the **Population** and **Year** fields to **Columns** and **Rows**, respectively.

Good structure – star schemas (data mart / data warehouse)

Assuming they are well designed, **star schema** data models work very well with Tableau because they have well-defined granularity, measures, and dimensions. Additionally, if they are implemented well, they can be extremely efficient to query. This leads to a good experience when using live connections in Tableau.

Star schemas are so named because they consist of a single fact table surrounded by related dimensions forming a star pattern. The following diagram illustrates a simple star schema with a single fact table (**Hospital Visit**) and three dimensions (**Physician**, **Treatment**, and **Patient**):

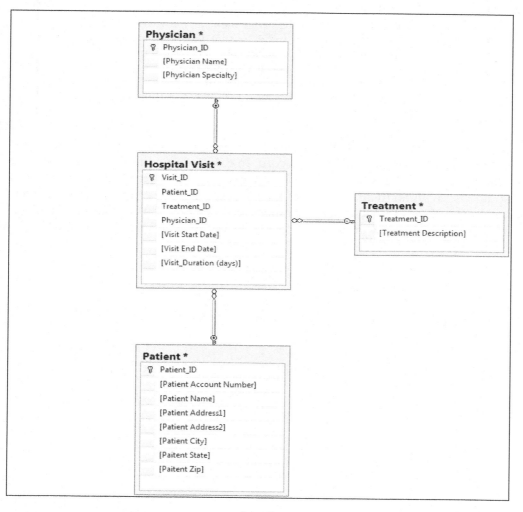

Fact tables are joined to the related dimension using what is often called a **surrogate key**, a foreign key that references a single-dimension record. The fact table defines the level of granularity and contains measures. In this case, **Hospital Visit** has a granularity of one record for each visit. Each visit, in this simple example, is for one patient who saw one physician and received one treatment. The **Hospital Visit** table explicitly stores a measure of **Visit Duration** and implicitly defines another measure of **Number of Visits** (that is, **Number of Records**).

Data modeling purists would point out that date values have been stored in the preceding fact table and would instead recommend that you have a date dimension table with extensive attributes for each date and only a surrogate (foreign) key stored in the fact table.

A date dimension can be very beneficial. However, Tableau's built-in date hierarchy and extensive date options make storing a date in the fact table a viable option. Consider using a date dimension if you need specific attributes of dates not available in Tableau (for example, days that are corporate holidays), have complex fiscal years, or if you need to support legacy BI reporting tools.

A well-designed star schema allows the use of inner joins, as every surrogate key should reference a single-dimension record. In cases where dimension values are not known or are not applicable, special dimension records are used. For example, a hospital visit with no treatment would have a surrogate key referencing a treatment record with a description such as "treatment not applicable." When connecting to a star schema in Tableau, start with the fact table and then add the dimension tables, as shown here:

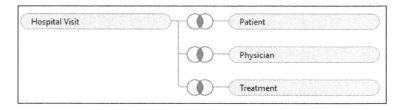

The resulting data connection allows you to see the dimensional attributes by table. The measures come from the single fact table.

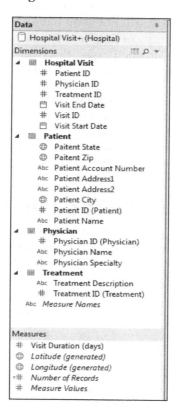

Well-implemented star schemas are particularly attractive for use in **Live** connections because Tableau can gain performance by implementing **join culling**.

Join culling is Tableau's elimination of unnecessary joins in queries it sends to the data source engine. For example, if you were to place **Physician Name** on **Rows** and the average of **Visit Duration** on **Columns** to get a bar chart of the average visit duration per physician, then joins to the **Treatment** and **Patient** tables may not be needed.

Tableau will eliminate unnecessary joins as long as you are using a simple star schema with only joins from the central fact table, have referential integrity enabled in the source, or allow Tableau to assume referential integrity (select the data source connection from the **Data** menu or use the context menu from the data source connection and choose **Assume Referential Integrity**).

Dealing with data structure issues

In some cases, restructuring data at the source is not an option. The source may be secured and read-only. Or, you might not even have access to the original data and might be receiving periodic dumps of data in a specific format instead, which would be tedious to change each time. In such cases, there are techniques to deal with structural issues once you have connected to the data in Tableau.

We'll consider some examples of data structure issues to demonstrate some techniques to handle these issues in Tableau. None of the solutions are the "only right way" to resolve the given issue. Often, there are several approaches that might work. Take time to understand how the proposed solutions build on the foundational principals we considered in previous chapters.

Restructuring data in Tableau connections

The `World Population Data.xlsx` Excel workbook, included in the `Data` directory of the resources included with this book, is typical of many Excel documents. Here is what it looks like:

	A	B	C	D	E	F	G	H	I	J	K
1	World Population Data										
3		This is sample data only.									
4		Accuracy and completeness is not guaranteed.									
5											
6	Country Name and Code	Indicator Name	Indicator Code	1960	1961	1962	1963	1964	1965	1966	1967
7	Aruba (ABW)	Population, total	SP.POP.TOTL	54208	55435	56226	56697	57029	57360	57712	58049
8	Andorra (AND)	Population, total	SP.POP.TOTL	13414	14376	15376	16410	17470	18551	19646	20755
9	Afghanistan (AFG)	Population, total	SP.POP.TOTL	8774440	8553544	9141783	9339507	9547131	9765015	9990125	10221902
10	Angola (AGO)	Population, total	SP.POP.TOTL	4965988	5056688	5150076	5245015	5339893	5433841	5526653	5619643
11	Albania (ALB)	Population, total	SP.POP.TOTL	1608800	1659800	1711319	1762621	1814135	1864791	1914573	1965598
12	United Arab Emirates (ARE)	Population, total	SP.POP.TOTL	89608	97727	108774	121574	134411	146341	156890	167360
13	Argentina (ARG)	Population, total	SP.POP.TOTL	20623998	20959241	21295290	21630854	21963952	22293817	22618887	22941477
14	Armenia (ARM)	Population, total	SP.POP.TOTL	1867396	1934239	2002170	2070427	2138133	2204650	2269475	2332624
15	American Samoa (ASM)	Population, total	SP.POP.TOTL	20012	20478	21118	21883	22701	23518	24320	25116
16	Antigua and Barbuda (ATG)	Population, total	SP.POP.TOTL	54681	55403	56311	57368	58500	59653	60818	62002
17	Australia (AUS)	Population, total	SP.POP.TOTL	10276477	10483000	10742000	10950000	11167000	11388000	11651000	11799000
18	Austria (AUT)	Population, total	SP.POP.TOTL	7047539	7086299	7129864	7175811	7223801	7270889	7322066	7376998
19	Azerbaijan (AZE)	Population, total	SP.POP.TOTL	3897889	4030130	4167558	4307315	4445653	4579759	4708485	4832098

Excel documents like this are often more human-readable but contain multiple issues for data analysis in Tableau. The issues in this particular document include:

- Excessive headers (titles, notes, and formatting) that are not part of the data
- Merged cells
- The country name and code in a single column
- Columns that are likely to be unnecessary (**Indicator Name** and **Indicator Code**)
- The data is wide: there is a column for each year and the population measure is spread across these columns within a single record

When we initially connect to the Excel document in Tableau, the connection screen will look similar to this:

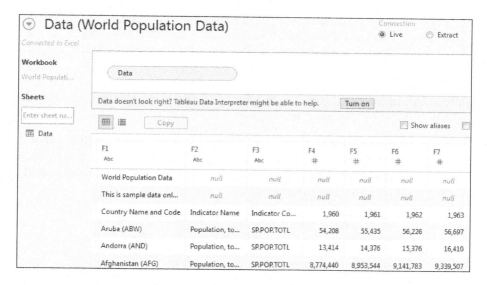

The data preview reveals some of the issues resulting from the poor structure:

- As the column headers were not in the first Excel row, Tableau gave the defaults of **F1**, **F2**, and so on to each column

- The title **World Population Data** and note about sample data were interpreted as values in column **F1**

- The actual column headers are treated as a row of data

Fortunately, these issues can be addressed in the connection window. First, we can correct many of the excessive header issues by turning on the **Tableau Data Interpreter**, a component that specifically identifies and resolves common structural issues in Excel documents. When you click on the **Turn on** button, the data preview reveals much better results.

Country Name and ...	Indicator Name	Indicator Code	1960	1961	1962	1963	1964	1965	1966
	Abc	Abc	#	#	#	#	#	#	#
Aruba (ABW)	Population, total	SP.POP.TOTL	54,208	55,435	56,226	56,697	57,029	57,360	57,712
Andorra (AND)	Population, total	SP.POP.TOTL	13,414	14,376	15,376	16,410	17,470	18,551	19,646
Afghanistan (AFG)	Population, total	SP.POP.TOTL	8,774,440	8,953,544	9,141,783	9,339,507	9,547,131	9,765,015	9,990,125
Angola (AGO)	Population, total	SP.POP.TOTL	4,965,988	5,056,688	5,150,076	5,245,015	5,339,893	5,433,841	5,526,653
Albania (ALB)	Population, total	SP.POP.TOTL	1,608,800	1,659,800	1,711,319	1,762,621	1,814,135	1,864,791	1,914,573

Clicking on the **Review results...** button will cause Tableau
to generate a new Excel document that is color-coded in
order to indicate how the Tableau Data Interpreter parsed
the Excel document. Use this feature to verify that Tableau
has correctly interpreted the Excel document and retained
the data you expect.

Observe the elimination of the excess headers and the correct names of columns.
A few additional issues still need to be corrected.

First, we can hide the **Indicator Name** and **Indicator Code** columns if we feel they
are not useful for our analysis. Clicking on the drop-down arrow on a column header
reveals a menu of options. **Hide** will remove the field from the connection and even
prevent it from being stored in extracts:

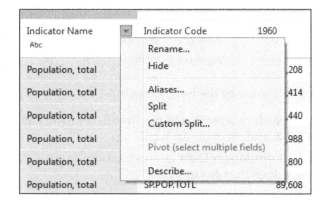

Second, we can split the **Country Name** and **Country Code** column into two
columns so that we can work with the name and code separately. In this case, the
Split option on the menu works well and Tableau perfectly splits the data, even
removing the parentheses from around the code. In cases where the split is not
automatically recognized, you may use the **Custom Split...** option. We'll also use the
Rename... option to rename the split fields from Country Name and Code - Split
1 and Country Name and Code - Split 2 to Country Name and Country Code,
respectively. Then, we'll hide the original Country Name and Code field.

At this point, most of the data structure issues have been remedied, with the exception of the wide nature of the data.

Country Name	Country Code	1960	1961	1962	1963	1964	1965
Andorra	AND	13,414	14,376	15,376	16,410	17,470	18,551
Afghanistan	AFG	8,774,440	8,953,544	9,141,783	9,339,507	9,547,131	9,765,015
Angola	AGO	4,965,988	5,056,688	5,150,076	5,245,015	5,339,893	5,433,841
Albania	ALB	1,608,800	1,659,800	1,711,319	1,762,621	1,814,135	1,864,791

Our final step is to **pivot** the **Year** columns. This means that we'll reshape the data in such a way that every country will have a row for every year. Select all the year columns by clicking on the area above 1960, scrolling to the far right, and hold *Shift* while clicking on the area above 2013. Finally, use the drop-down menu on any one of the year fields and select **Pivot**.

The result is two columns (`Pivot field names` and `Pivot field values`) in place of all the year columns. Rename the two new columns to `Year` and `Population`. Your dataset is now narrow and tall instead of wide and short. This dataset will be far easier to work with in Tableau than the original.

Country Name	Country Code	Year	Population
Aruba	ABW	1960	54,208
Andorra	AND	1960	13,414
Afghanistan	AFG	1960	8,774,440
Angola	AGO	1960	4,965,988
Albania	ALB	1960	1,608,800
United Arab Emirates	ARE	1960	89,608
Argentina	ARG	1960	20,623,998
Armenia	ARM	1960	1,867,396
American Samoa	ASM	1960	20,012
Antigua and Barbuda	ATG	1960	54,681
Australia	AUS	1960	10,276,477
Austria	AUT	1960	7,047,539

> Tableau 9.0 introduced pivots for Excel and text files. Previous versions of Tableau do not have the built-in pivot feature. However, there is a free reshaping add-in for Excel made available by Tableau. This knowledge base article describes the tool and includes a link for download at `http://kb.tableausoftware.com/articles/knowledgebase/addin-reshaping-data-excel`.

Working with poorly shaped data in visualizations

If possible, it is best to work with data that is structured well or reshape the data prior to or while connecting to it in Tableau. But what if you had wide data that couldn't be fixed at the source? While the pivot option is available for text and Excel files, it isn't available for other data sources. Perhaps your data is in a read-only table in your corporate database or maybe you only have access to an extract. Let's say you have been given a wide table of data like this with no option to reshape it at the source:

Country name	1960	1961	1962	1963	1964
Afghanistan	8,774,440	8,953,544	9,141,783	9,339,507	9,547,131
Australia	10,276,477	10,483,000	10,742,000	10,950,000	11,167,000

You would find this data difficult to work with in Tableau. Because each year is treated as a separate measure, you might start by dragging and dropping each field into the view. It would quickly become obvious that separate measures placed on **Rows** or **Columns** are not at all what you want. You end up with different axes or layers of headers.

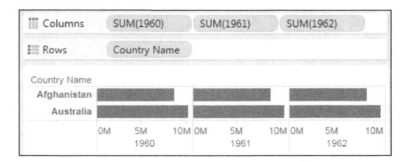

Although it is difficult to use, the poorly structured data is not impossible to use. You'll recall encountering the special **Measure Names** and **Measure Values** fields in *Chapter 3, Moving from Foundational to Advanced Visualizations*. **Measure Names** serves as a dimensional placeholder for the names of all measures and **Measure Values** serves as a placeholder for all the values of the measure fields.

Using these special fields allows you to build views such as this:

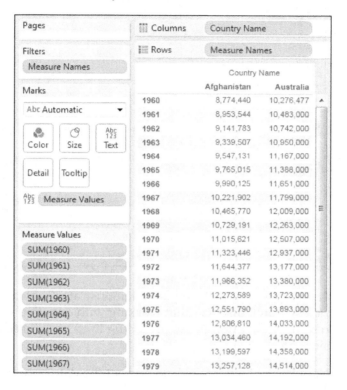

Note how the **Measure Names** field on **Rows** gives a row for each measure on the **Measure Values** shelf and **Measure Values** on **Text** tells Tableau to draw the text value of each measure as a mark in the pane.

> Rather than dragging and dropping each measure to the **Measure Values** shelf individually, click on the first measure in the **Data** window, hold the *Shift* key, and then click on the last measure to select everything between the first and last measure. You may then drag the entire group of measures to the shelf. (You can also use the *Ctrl* key to individually select measures and then drag and drop as a group.)

You can even rearrange the fields to create useful visualizations similar to those possible with tall data. For example, you can create a line chart by moving the **Measure Names**, **Measure Values**, and **Country Name** fields and changing the mark type to **Line**. The result is a view like this:

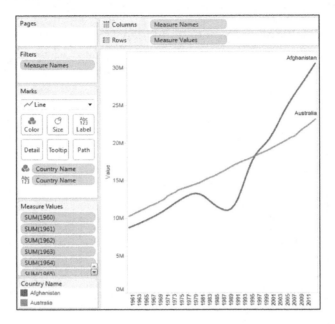

The ability to work with poorly structured data demonstrates Tableau's impressive flexibility. Nevertheless, you'll discover that working with data that is not optimally structured leads to some difficulties. In the preceding line chart, these are some of the limitations we'd encounter:

- **Measure Names** is not a date field, so none of the date options or formats can be applied.
- Without an actual date, we are not able to use trend lines or forecasting.
- **Measure Names** is discrete, so each year is a header. We cannot change it to continuous to get an axis.
- The **Measure Names** and **Measure Values** fields cannot be used in calculations, sets, groups, or bins.
- Quick table calculations cannot be applied to **Measure Values**, so getting a running total or percent difference would not be possible. It is possible to apply a quick table calculation to the individual measures, but they would be meaningless. Furthermore, although **Measure Names** functions as a dimension in some ways, it cannot be used for the addressing or partitioning of table calculations.

Some of the limitations listed here can be addressed using additional techniques mentioned in the upcoming *An overview of advanced fixes for data problems* section.

Working with an incorrect level of detail

Remember that the two keys of good structure are having a level of detail that is meaningful and having measures that match the level of detail or that are possibly at higher levels of detail. Measures at lower levels tend to result in wide data. Measures at higher levels of detail can, at times, be useful and as long as we are aware of how to handle them, we can avoid some pitfalls.

Consider, for example, this data that gives us a single record each month per apartment:

Apartment	Month of Month	Rent Collected	Square Feet
A	January 2014	$0	900
	February 2014	$0	900
	March 2014	$0	900
	April 2014	$0	900
	May 2014	$0	900
	June 2014	$0	900
	July 2014	$1,500	900
	August 2014	$1,500	900
	September 2014	$1,500	900
	October 2014	$1,500	900
	November 2014	$1,500	900
	December 2014	$1,500	900
B	January 2014	$1,200	750
	February 2014	$1,200	750
	March 2014	$1,200	750
	April 2014	$1,200	750
	May 2014	$1,200	750
	June 2014	$1,200	750
	July 2014	$0	750
	August 2014	$0	750
	September 2014	$0	750
	October 2014	$0	750
	November 2014	$0	750
	December 2014	$0	750

The two measures are really at different levels of detail:

- **Rent Collected** matches the level of detail of the data; there is a record of how much rent was collected for each apartment for each month
- **Square Feet**, on the other hand, does not change month to month. Rather, it is at the higher level of apartment

This can be observed when we remove the date from the view and look at everything at the apartment level:

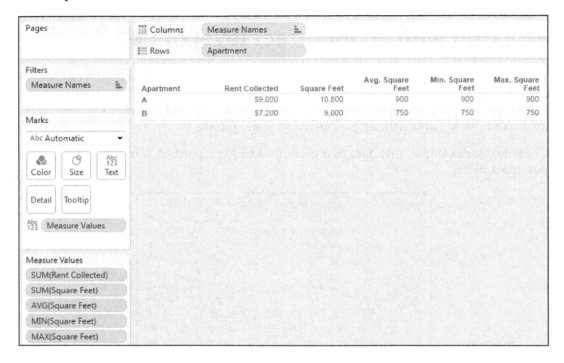

Note that the **Sum** value of **Rent Collected** makes perfect sense. You can add up the rent collected per month and get a meaningful result per apartment. However, you cannot sum up **Square Feet** and get a meaningful result per apartment. Other aggregations, such as the average, minimum, and maximum, do give the right results per apartment.

However, imagine that you were asked to come up with the ratio of total rent collected per square feet per apartment. You know this will be an aggregate calculation because you have to sum the rent collected. But which of these is the correct calculation? Take a look:

- SUM([Rent Collected]) / SUM([Square Feet])
- SUM([Rent Collected]) / AVG([Square Feet])
- SUM([Rent Collected]) / MIN([Square Feet])
- SUM([Rent Collected]) / MAX([Square Feet])

The first one is obviously wrong. We've already seen that square feet should not be added each month. Any of the final three would be correct if we ensure the apartment continues to define the level of detail of the view. However, once we look at the view that has a higher level of detail (for example, multiple or all apartments), the calculations don't work. To understand why, consider what happens when we turn on column grand totals (from the menu, navigate to **Analysis | Totals | Show Column Grand Totals** or drag and drop **Totals** from the **Analytics** tab).

Apartment	Rent Collected	Square Feet	Avg. Square Feet	Min. Square Feet	Max. Square Feet
A	$9,000	10,800	900	900	900
B	$7,200	9,000	750	750	750
Grand Total	$16,200	19,800	825	750	900

The problem here is that the **Grand Total** line is at the level of detail of "all apartments (for all months)". What we really want as the grand total of square feet is 900 + 750 = 1,650. But here, the sum of square feet is the addition of square feet for all apartments for all months. The average is overall. The minimum finds the value 750 as the smallest measure for all apartments in the data. The maximum, likewise, picks the single largest value. So at any level of detail higher than individual apartments, none of the proposed calculations would work.

You can adjust how subtotals and grand totals are computed by clicking on the individual value and using the drop-down menu to select how the total is computed. Alternately, right-click on the active measure field and select **Total Using**.

You can change how all measures are totaled at once from the menu by navigating to **Analysis | Totals | Total All Using**. Using this **two pass total** technique could result in correct results in the preceding view but would not solve the issue of needing the correct total in another calculated field.

Fortunately, Tableau gives us the ability to work with different levels of detail in a view. Using **level of detail (LOD)** calculations, which we encountered briefly in *Chapter 4, Using Row-level and Aggregate Calculations*, we can calculate the square feet per apartment.

Here, we'll use a fixed LOD calculation to keep the LOD fixed at the apartment. We'll create a calculated field named Square Feet per Apartment with this code:

```
{ FIXED [Apartment] : MIN([Square Feet]) }
```

The curly braces surround an LOD calculation and the FIXED keyword indicates that we want the following aggregation to always be performed at the LOD of the dimension(s) following FIXED, regardless of what level of detail is defined in the view. Here, we decided to get the minimum square feet per apartment. As the area doesn't change in our data, we could have used maximum or average. The result is a calculated field that always gives us the square feet per apartment.

As you can see, the calculation returns the correct result in the view at the apartment level and at the grand total level:

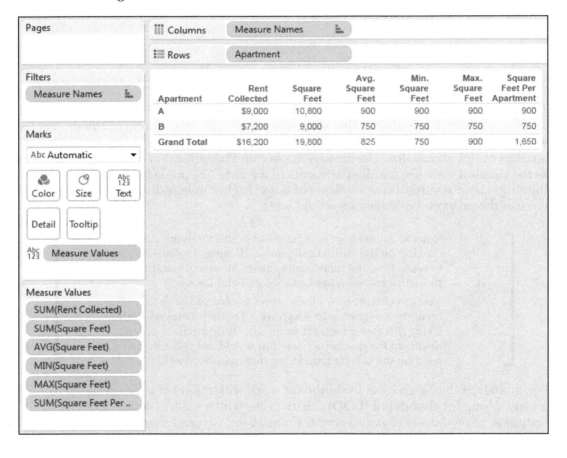

Now, we can use the LOD calculated field in another calculation to determine the desired results. We'll create a calculated field named Rent Collected per Square Foot with the code:

```
SUM([Rent Collected]) / SUM([Square Feet Per Apartment])
```

When that field is added to the view, the final outcome is correct:

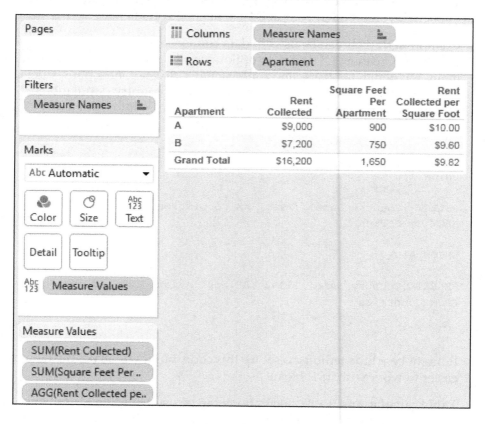

An overview of advanced fixes for data problems

In addition to the techniques shown in the previous sections, there are some additional possibilities for dealing with data structure issues. It is outside the scope of this book to develop these concepts fully. However, given some familiarity with these approaches, you broaden your ability to deal with challenges as they arise:

- **Custom SQL**: This can be used in data connection to resolve some data problems. Custom SQL is not an option for all data sources but is for many relational databases and for legacy JET driver connections. Consider a custom SQL script that takes the wide table of country populations mentioned earlier in this chapter and restructures it into a tall table:

```
SELECT [Country Name],[1960] AS Population, 1960 AS Year
FROM Countries

UNION ALL

SELECT [Country Name],[1961] AS Population, 1961 AS Year
FROM Countries

UNION ALL

SELECT [Country Name],[1962] AS Population, 1962 AS Year
FROM Countries
...
...
```

  It might be a little tedious to set up this code, but it will make the data much easier to work with in Tableau!

- **Table calculations**: Table calculations can be used to solve a number of data challenges, from finding and eliminating duplicate records to working with multiple levels of detail. As table calculations can work within partitions at higher levels of detail, you can use multiple table calculations and aggregate calculations together to mix levels of detail in a single view. A simple example of this is the **Percent of Total** table calculation, which compares an aggregate calculation at the level of detail in the view with a total at a higher level of detail.

- **Data blending**: Data blending can be used to solve numerous data structure issues. Because you can define the linking fields used, you control the level of detail of the blend. For example, the preceding rental data problem could be solved with a secondary source that had a single record per apartment with the square feet. Blending at the apartment level would allow you to achieve the desired results.

- **Data scaffolding**: Data scaffolding extends the concept of data blending. With this approach, you construct a "scaffold" of various dimensional values to use as a primary source and then blend to one or more secondary sources. This way, you can control the structure and granularity of the primary source while still being able to leverage data contained in secondary sources.

Summary

Up until this chapter, we looked at data that was, for the most part, well structured and easy to use. In this chapter, we considered what constitutes good structure and how to deal with poor data structure. Good structure consists of data that has a meaningful level of detail and that has measures that match that level of detail. When measures are spread across multiple columns, we get data that is *wide* instead of *tall*. You've got some experience now in applying various techniques to deal with data that has a wrong shape or has measures at the wrong level of detail. Tableau gives you the power and flexibility to deal with these structural issues, but it is far preferable to fix the data structure at the source.

In the next chapter, we'll continue looking at some advanced and powerful techniques. These will be exciting and fun. Instead of looking at how to fix problems, we'll look at some tips and tricks to expand your creativity and take Tableau to the next level!

10
Advanced Techniques, Tips, and Tricks

With a solid understanding of the foundational principles, it is possible to push the limits with Tableau. In addition to exploring, discovering, analyzing, and communicating data, members of the Tableau community have used the software to create and do amazing things such as simulate an enigma machine, play tic tac toe or blackjack, generate fractals with only two records of data, and much more! Unlike traditional BI packages that force you to go through a series of wizards to create a chart based on a predefined template, Tableau really is a blank canvas and the only limits are your creativity and imagination.

In this chapter, we'll take a look at some advanced techniques in a practical context. You'll learn things such as dynamically swapping views on a dashboard, using custom images, and advanced geographic visualizations. The goal of this chapter is not to provide a comprehensive list of every possible technique. Instead, we'll take a look at a few varied examples that demonstrate some possibilities. Many of the examples are designed to stretch your knowledge and challenge you.

We'll take a look at the following advanced techniques in this chapter:

- Sheet swapping and dynamic dashboards
- Leveraging sets to answer complex questions
- Advanced mapping techniques
- Using background images

Sheet swapping and dynamic dashboards

Sheet swapping, sometimes also called **sheet selection,** is a technique in which views are dynamically shown and hidden on a dashboard, often with the appearance of swapping one view for another. The dynamic hiding and showing of views on a dashboard has an even broader application. When combined with floating objects and layout containers, this technique allows you to create rich and dynamic dashboards.

The basic principles are simple:

- A view "collapses" on a dashboard when at least one field is in **Rows** or **Columns** and **Filters** and/or hiding prevent any marks from being rendered
- Titles and captions do not collapse but can be hidden so that the view collapses entirely

Let's consider a simple example with a view showing **Profit by Department and Category** with a **Department** quick filter. The dashboard has been formatted (from the menu, navigate to **Format | Dashboard**) with a blue shading to help us see the effect:

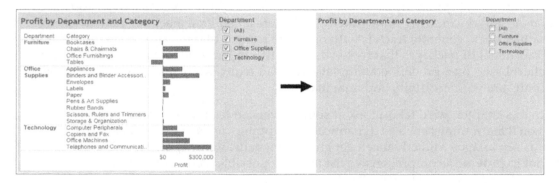

Observe how filtering out all the departments results in the view collapsing. The title remains, but it could have been hidden.

In order to swap two different sheets, we simply take advantage of the collapsing behavior along with the properties of layout containers. We'll start by creating two different views filtered via a parameter and a calculated field. The parameter will allow us to determine which sheet is shown. Perform the following steps:

1. Create an **Integer** parameter named Show Sheet with a list of values consisting of 1 and 2 and with the **Display As** text set to Bar Chart and Map respectively:

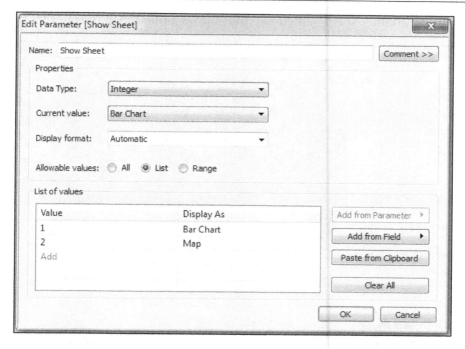

2. We want to filter based on the parameter selection, and the parameters cannot be directly added to the **Filters** shelf; instead, we'll create a calculated field named Show Sheet Filter to return the selected value of the parameter. The code is simply [Show Sheet].

3. Drag this new calculated field to the **Dimensions** section of the **Data** window. This is not necessary but will make it easier to filter.

4. Create a new sheet named Bar Chart that is similar to the **Profit by Department** and **Profit by Category** view, as shown in the preceding screenshot.

5. Show the parameter control (right-click on the parameter in the **Data** window and select **Show Parameter Control**). Make sure the **Bar Chart** option is selected.

6. Add the **Show Sheet Filter** field to the **Filters** shelf and check the value **1 (Bar Chart)** to keep.

7. Create another sheet named Map that shows a filled map of states by profit:

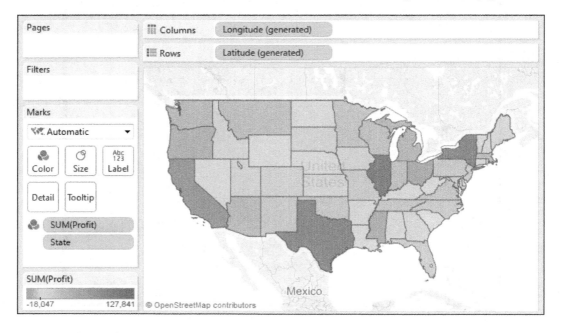

8. Show the parameter in this view and change the selection to **Map**. Remember that the parameter selections are universal to the worksheet. If you were to switch back to the bar chart view, it should no longer be showing any data because of the filter.

9. Add the **Show Sheet Filter** field to the **Filters** shelf and check the value **2** (**Map**) to keep.

10. Create a new dashboard named Sheet Swap.

11. Add a vertical layout container to the dashboard from the objects on the left-hand side of the window:

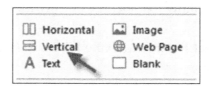

A horizontal layout container would work just as well in this case. The key is that a layout container will allow each view inside to expand to fill the container when the view is set to fit entire view, fit width (for horizontal containers), or fit height (for vertical containers). When one view collapses, the visible view will expand to fill the rest of the container.

12. Add each sheet to the layout container in the dashboard. The parameter control should be added automatically since it was visible in each view.

13. Using the drop-down menu in the **Bar Chart** view, navigate to and set the view to **Fit | Entire View**. Geographic visualizations automatically fill the container.

14. Hide the title for each view.

15. You now have a dashboard where changing the parameter results in one view or the other being shown. When **Map** is selected, the filter results in no data for the bar chart, so it collapses and **Map** fills the container:

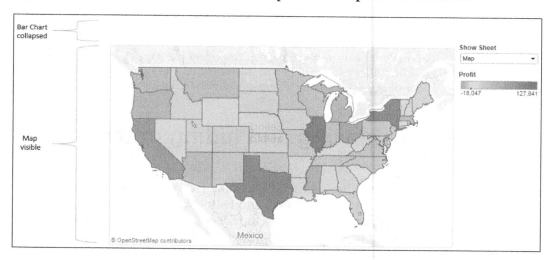

16. Alternately, when **Bar Chart** is selected, the **map** collapses due to the filter and the bar chart fills the container:

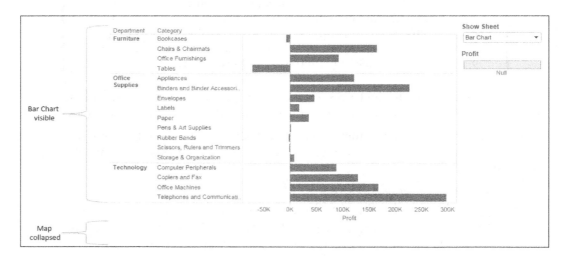

The key to collapsing a view is to have a filter or set of filters that ensures no rows of data. You do not have to use a parameter to control the filtering. You could use a quick filter or an action filter to accomplish the same effect. This opens up all kinds of possibilities for dynamic behavior in dashboards.

Dynamically showing and hiding other controls

Views will collapse based on filtering out all the data. However, other controls, such as quick filters, parameters, images, legends, and textboxes, will not collapse and don't have an option to dynamically show or hide. Yet many times in a dynamic dashboard, you might want to show or hide these objects. Sometimes, parameters don't apply when other selections have been made. Take a look at the preceding example. The color legend that was automatically added to the dashboard by Tableau applies to the map. However, when the bar chart is shown, the legend is no longer applicable.

Fortunately, we can extend the technique we used to collapse or expand views that "push" objects from behind floating objects to make them appear or disappear on a dashboard.

Let's extend the preceding example to demonstrate how to show and hide the color legend:

1. Create a new sheet named `Show/Hide Legend`. This view is only used to show and hide the color legend.

2. Create a calculated field named `One` with the code of `1`. We must have a field in **Rows** or **Columns** for the view to collapse, so we'll use this field to give us a single axis for **Rows** and **Columns** without any other headers.

3. Add the **One** field to **Rows** and **Columns**. You should now have a simple scatterplot with one mark.

4. As this is a helper sheet and something we want the user to see, we don't want it to show any marks or lines. Format the view by navigating to **Format | Lines** to remove **Grid Lines**, and navigate to **Format | Borders** to remove **Row Dividers** and **Column Dividers**. Additionally, hide the axes (right-click on each axis or field and uncheck **Show Headers**). Also, set **Color** to **Full Transparency** to hide the mark.

5. We will want this view to show when the **Map** option is selected, so show the parameter control and ensure that it is set to **Map**. Then, add the **Show Sheet Filter** to filters and check the value **2** (**Map**):

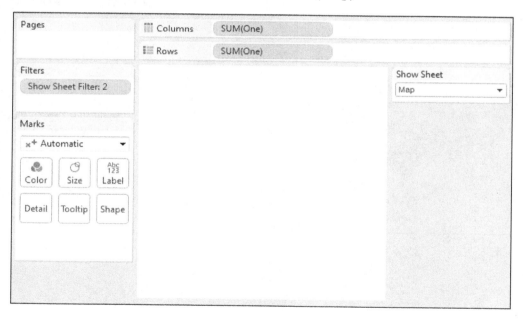

6. On the sheet's **Swap Dashboard**, add the **Show/Hide Legend** sheet to the layout container between the **Show Sheet** parameter dropdown and the color legend.

7. Ensure that **Map** is selected. The color legend should be pushed all the way to the bottom.

8. Add a layout container as a floating object. Size and position it to completely cover the area where the color legend used to be. It should cover the title of the **Show/Hide Legend** sheet but not the parameter dropdown.

 Objects and sheets can be added as floating objects by holding *Shift* while dragging, setting the **New Objects** option to **Floating**, or using the drop-down menu on the object.

9. The layout container is transparent by default, but we want it to hide what is underneath. Format the layout container using the drop-down menu, and add white shading so that it is indistinguishable from the background.

10. At this point, you have a dynamic dashboard where the legend is shown when the map is shown, and it is applicable and hidden when the bar chart is visible. When **Map** is selected, the **Show/Hide Legend** sheet is shown, and it pushes the legend to the bottom of the layout container:

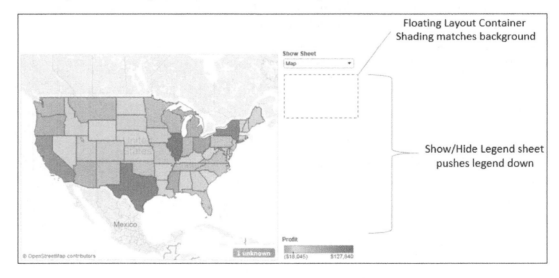

11. When **Bar Chart** is selected, the **Show/Hide Legend** sheet collapses and the legend, which is no longer applicable to the view, falls under and hides behind the floating layout container:

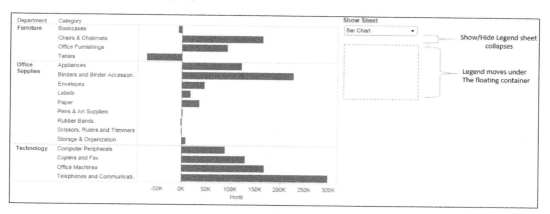

There is no limit to how many variations of this technique you can use on a dashboard. You can have as many layers as you'd like. The possibilities to create a richly interactive user experience are endless.

Leveraging sets to answer complex questions

Sets are an extremely powerful feature of Tableau. They enable valuable analysis and can be used to answer incredibly complex questions.

Sets are special fields in Tableau that define a grouping of data by effectively identifying each underlying record of data as in or out of the set based on certain conditions you define.

Sets can be created in a couple of ways:

- Select one or more marks in a view, and use the **Create Set** option on the tooltip controls. The set will contain every combination of dimension values that made up the selected marks:

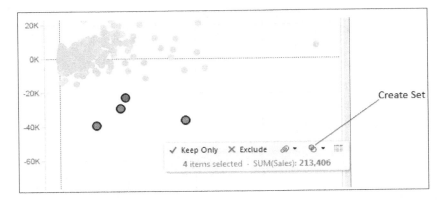

- Create a set from a single dimension by right-clicking on the **Dimension** field in the **Data** window and selecting **Create Set...** The resulting dialog box gives you the option of selecting individual dimension members (under the **General** tab) or setting specific conditions that define whether the data belongs to the set (under the **Condition** or **Top** tab):

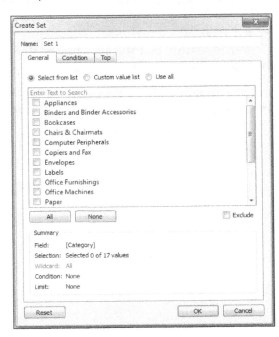

Sets that are created from marks or by selecting individual dimension values may be edited but remain static. That is, they will always be defined by the combination of certain dimension values. Sets that are defined using **Condition** or **Top** are called **calculated sets** or **computed sets** and are computed as part of the query of the source data.

Sets appear in their own section as fields in the **Data** window. You can drag these set fields onto shelves in the view. The set field can use one of two modes that can be changed using the drop-down menu in the field:

- **Show In/Out of Set**: This field acts as a dimension that slices the data based on whether the underlying records are in or out of the set. This option is not available when using legacy JET connections.

- **Show Members in Set**: A filter is added to keep only the members of the set, and the field acts as a dimension that slices the data for each member of the set.

Answering complex questions

Sets are useful in many ways. For example, you can visually define a set by selecting a group of marks on a scatterplot that represents unprofitable items with high sales. Then, you can further do visual analysis on that set or compare it to other sets. Additionally, since sets allow you to define a subset of data, you can define the set once, and the set becomes an easy reusable filter.

Beyond these uses, sets provide a tool for answering extremely complex questions. Let's say, for example, you are working for the superstore management and they ask the following question:

- Compare the overall 2015 profit (for all departments) for customers who purchased only from the furniture department in 2014 versus customers who purchased only from the technology department in 2014

These types of questions seem almost impossible to answer, short of writing a complex script. Even in Tableau, it would be challenging to use the techniques we've looked at so far. If you filter only by the year 2014, you won't be able to see the 2015 profit. Filtering by either furniture or technology excludes the other and doesn't help answer whether the customer *only* purchased from that department. It might be possible to answer this question using table calculations, though the solution would be extremely complex.

Another, simpler, option is to use sets. Looking at the question, we can identify two distinct sets:

- Customers who purchased only furniture in 2014
- Customers who purchased only technology in 2014

Let's take a look at how we can create and leverage these sets to achieve a solution:

1. Create a calculated field named `2014 Department` with the following code:

   ```
   IF YEAR([Order Date]) = 2014 THEN [Department] END
   ```

   This is a row-level calculation that will return the department if the year of the order was 2014 and NULL otherwise. We'll use this field to determine whether a customer ordered from only one department and whether it was furniture or technology.

2. Right-click on the **Customer Name** field and select **Create Set**. Name the set `Only Furniture in 2014.`

    This example uses **Customer Name**, which we know represents a unique customer in the sample dataset. In practice, consider using fields such as **Customer ID** instead of **Customer Name**. ID fields should always be unique, but it is conceivable that multiple individuals might share the same name.

3. Use the **Condition** tab and select **By formula**. This formula will be an aggregate formula performed at the level of a customer. So, for each customer, we will write the formula that will return `true` (if they are in the set) or `false` (if they are out of the set). Here, we'll use the following code:

   ```
   MIN([2014 Department]) = "Furniture"
   AND
   MAX([2014 Department]) = "Furniture"
   ```

This is the code in the textbox of the following screenshot:

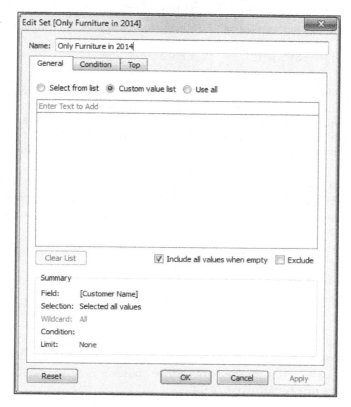

The set will be calculated across the entire set of the source data for each customer. This means that if, in 2014, a customer's minimum department was **Furniture** and the maximum was also **Furniture**, then that customer only purchased from the **Furniture** department in that year.

4. Create another, similar set, named `Only Technology in 2014`. Again, define the set by selecting **Condition** and **By formula** with similar code:

```
MIN([2014 Department]) = "Technology"
AND
MAX([2014 Department]) = "Technology"
```

> Although we won't use the feature in this example, Tableau does allow you to create combined sets. **Combined sets** allow you to take two sets and determine which members of the sets are shared or exclusive to each set. To create a combined set, select two sets in the **Data** window, right-click, and select **Create Combined Set**.

5. In order to compare the profit for members of each set, we'll create a row-level calculated field that will slice the data based on which set identifies a record. Create a calculated field called `Customer Set` with the following code:

```
IF [Only Furniture in 2014]
    THEN "Customers who only ordered furniture in 2014"
ELSEIF [Only Technology in 2014]
    THEN "Customers who only ordered technology in 2014"
ELSE "Neither technology nor furniture in 2014"
END
```

6. Create a view that uses the **Customer Set** field on **Rows** to slice the data. We'll include **Profit** in **Columns**, and then filter the order date to the year 2015; we'll also exclude the row for the customers who ordered neither technology nor furniture in 2014.

Our final view shows us that customers who ordered only technology in 2014 gave us slightly more profit than customers who ordered only furniture:

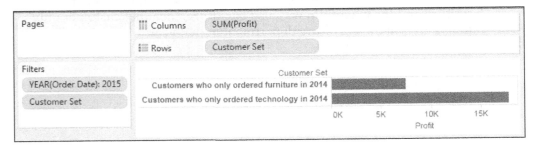

> Sets are computed within context. **Context** is either the entire underlying dataset or a subset of data defined by context filters. To add a filter to the context, simply right-click on it and select **Add to Context**.

Consider the power of sets in the preceding example. We were able to answer an incredibly complex question using just a couple of calculations and calculated sets. Tableau gave us some features that made this relatively easy compared to other options.

Consider the underlying complexity: our first calculation was calculated at a row level and then aggregated in the set calculations, and then another calculation was applied to each row of data to determine to which set it belonged. It will take some practice to be able to leverage sets to their full potential. The kinds of questions you will be able to answer will make it worthwhile!

Mapping techniques

We've touched on geographic visualization throughout the book. You've seen symbol maps and filled maps. Here, we'll examine an example using a custom geocoding technique and custom shapes to give some idea of what is possible.

Supplementing the standard geographic data

We saw in *Chapter 1, Creating Your First Visualizations and Dashboard*, that Tableau generates the **Latitude** and **Longitude** fields when the data source contains geographic fields, which Tableau can match with its internal geographic database. Fields such as country, state, zip code, MSA, and congressional district are contained in Tableau's internal geography.

However, if you have latitude and longitude in your dataset or are able to supplement your data source with that data, you can create geographic visualizations with great precision. There are several options for supplying latitude and longitude for use in Tableau:

- Include latitude and longitude as fields in your data source. If possible, this option will provide you with the easiest approach to create custom geographic visualizations because you can simply place **Latitude** in **Rows** and **Longitude** in **Columns** to get a geographic plot.

- Assign a field to a geographic role, and assign unrecognized values to the desired latitude and longitude locations. This option is most often used to correct unrecognized locations in standard geographic fields, such as a city or zip code but can be used to create custom geography as well, though it would be tedious and difficult to maintain more than a few locations.

You can assign unknown locations by clicking on the **unknown** indicator in the lower-right corner. This will give you the option to **Edit Locations**, filter out unknown locations, or plot at the default location (latitude and longitude of 0; this is a location that is sometimes humorously referred to as *Null Island*, which is located just off the west coast of Africa).

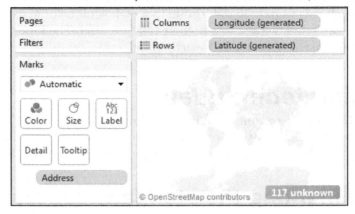

The **Edit Locations** option allows you to correct unmatched locations by selecting a known location or entering your own latitude and longitude information:

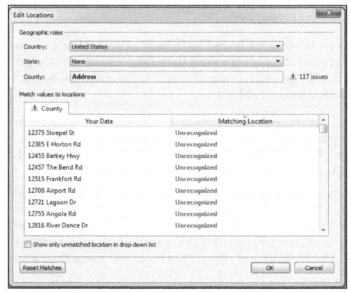

- Create a calculated field for latitude and another field for longitude using the IF/THEN logic or the case statement to assign the latitude and longitude values based on other values in your data. This would also be tedious and difficult to maintain over many locations.

- Import a custom geographic file. From the menu, navigate to **Map | Geocoding | Import Custom Geocoding…**. The import dialog box contains a link to the documentation that describes the option in more detail.

- Connect to the data that contains your latitudes and longitudes as a secondary data source, and use data blending to achieve geographic visualization.

Customizing a geographic view

Let's consider the final option as an example. As you can remember, we looked at the Real Estate data in *Chapter 8, Adding Value to Analysis – Trends, Distributions, and Forecasting*, when we discussed the statistical analysis. This data includes an **Address** field but does not contain any standard geographic fields such as city or state.

Data can be geocoded using a variety of tools and third-party providers. Often, the result is a data file that contains an identifier along with latitudes and longitudes, as shown in the following screenshot:

ID	Latitude	Longitude
1	29.88669	-81.63196
2	29.88669	-81.33866
3	29.89007	-81.33866
4	29.83001	-81.31622
5	29.76155	-81.29166
6	26.47224	-80.08122
7	29.92098	-81.35385
8	29.85276	-81.28601
9	29.96227	-81.53588
10	29.88669	-81.33939

Often, you will be able to include this supplemental data along with the original data by joining it together. However, if this is not possible, you can use Tableau to blend the original data with the supplemental data. The following steps use the **Real Estate** connection as the primary data source and **Real Estate (Supplemental)** as a secondary data source (as shown in the previous screenshot). The two data sources have an **ID** field in common:

1. Drag and drop the **Address** field from **Dimensions** to the **Marks** card. This will make **Real Estate** the primary connection, and define **Address** as the level of detail for the view. Ultimately, we want one mark per address.

2. Switch to the **Real Estate (Supplement)** data connection and drag **Latitude** to **Rows** and **Longitude** to **Columns**. Notice that Tableau recognizes latitude and longitude as geographic fields that together define a map. Using the **Address** field as a linking field, you should now see the 117 marks at the proper latitude and longitude:

3. Let's focus on houses built in 2000. Use the **Year Built** field in the primary data source as a filter.

4. By default, Tableau minimizes anything on the map that might distract you from understanding the data. It uses a gray map by default and doesn't show the various artifacts such as roads or county borders. However, sometimes the context can be helpful. In this case, from the menu, navigate to **Map | Options**, and then in the **Map Options** window, set the **Style** field to **Normal** and check **Streets and Highways**:

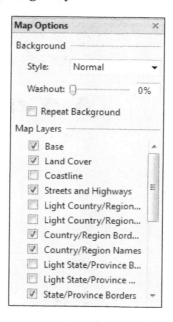

5. We'll use a custom shape to identify the type of each house. First, locate the three image files included in the `Learning Tableau\Custom Shapes\Houses` directory and copy them to the `Documents\Tableau Repository\Shapes\Houses` directory (create the `Houses` subdirectory if needed).

> You can add all the libraries of custom shapes to the `Documents\My Tableau Repository\Shapes` directory. Each subdirectory will appear as a category of shapes available to assign discrete values of fields to the **Shape** shelf.
>
> Shapes that are used in a workbook will be included in a packaged workbook and also if a workbook is published to the Server, Online, or Public.
>
> Higher resolution images will scale better, and Tableau will keep any transparency in the image.

6. Change the mark type to **Shape**.

7. Drag and drop the **Type** field to the **Shape** shelf. Edit the shape legend using the drop-down menu or by double-clicking on the legend, and assign each dimension value to the corresponding image. If the **houses** shapes category does not appear in the dropdown, click on **Reload Shapes**. This will cause Tableau to refresh the shapes available based on the contents of the directory:

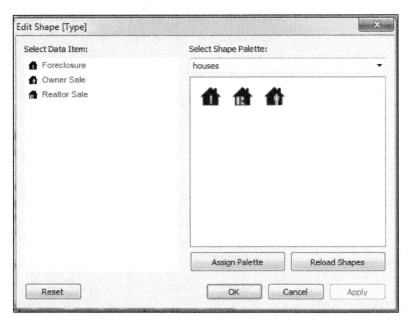

8. Drag and drop the **Area** measure to **Size**. Click on the **Size** shelf button, and adjust the slider as needed. You can also adjust the sizes by selecting **Edit** on the drop-down menu of the size legend.

9. Drag and drop the **Price** measure to the **Label** shelf.

You now have a geographic visualization using custom shapes at the exact location of each house for sale. Adjust the map options to give a map with enough detail to understand the context of each location:

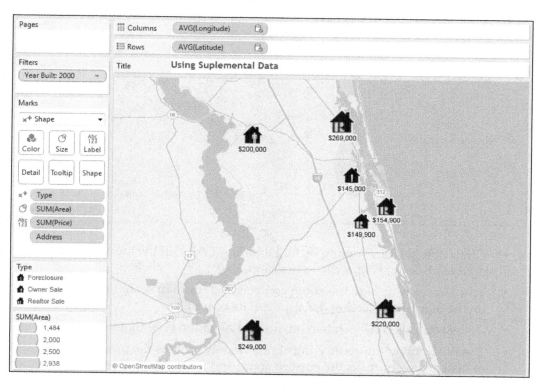

Concrete shapes can dramatically reduce the amount of time taken to comprehend the data. Contrast the amount of effort required to identify the departments in these two scatter plots:

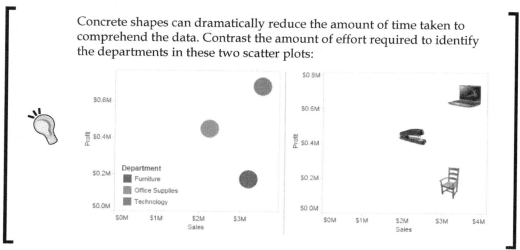

Some final mapping tips

Here are some final tips to keep in mind when creating geographic visualizations:

- Various controls will appear when you hover over the map. These will enable various zoom and selection options:

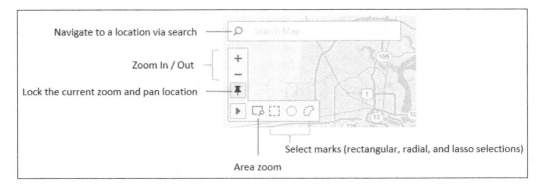

- Other options for zooming include holding *Ctrl* with the mouse wheel, double-clicking, *Shift* + *Alt* + click, and *Shift* + *Ctrl* + click.

- You can show or hide the zoom controls and/or map search by right-clicking on the map and selecting the appropriate option.

- Zoom controls can be shown on any visualization type that uses an axis.

- The pushpin on the zoom controls alternately returns the map to the best fit of visible data or locks the current zoom and location.

- You can create a dual axis map by duplicating (*Ctrl* + drag/drop) either the **Latitude** field in **Columns** or **Longitude** in **Rows**, and then using the field's drop-down menu to select **Dual Axis**. You can use this technique to combine multiple mark types on a single map:

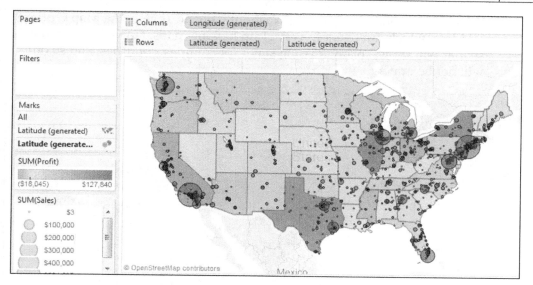

- You can plot pie charts on a map, which is one way to make comparisons of a parts-to-whole relationship, while simultaneously understanding geographic locations. Simply create a geographic visualization and change the mark type to **Pie**:

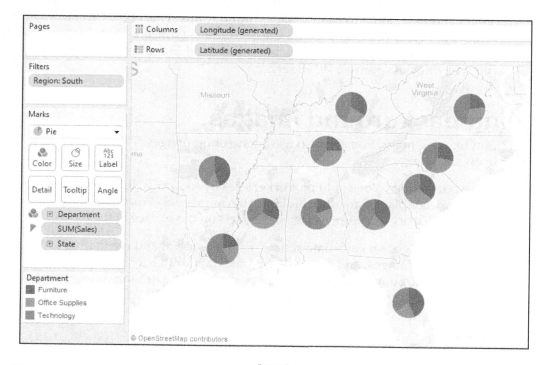

- When using filled maps, set the **Washout** value to `100%` in the **Map Options** window, which can result in very clean looking maps. However, only filled shapes will show, so any missing states (or counties, countries, and so on) will not be drawn:

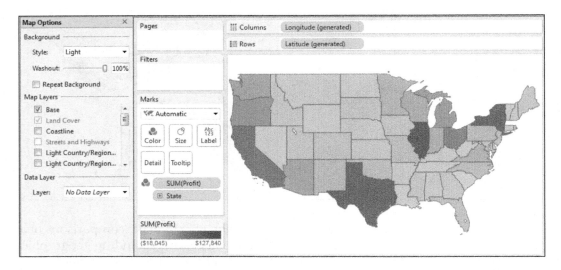

- You can change the source of the background map image tiles using the menu and navigating to **Map | Background Maps**. This allows you to choose between Tableau, Tableau Classic, Offline (which is useful when you don't have an Internet connection, but it is limited in the detail that can be shown), or even specify a WMS server.

Using background images

In addition to using maps, you can also use background images. This allows you to plot data on any image.

Consider the possibilities. You could plot ticket sales by seat on an image of a stadium, room use on the floor plan of an office building, the number of errors by a piece of equipment on a network diagram, or meteor impacts on the surface of the moon.

You may specify images for each data source using the menu and navigating to **Map | Background Images**, and then selecting the data source for which the image applies. On the **Background Images** screen, you can add one or more images that will be shown as the background of scatterplots.

When you add or edit an image, you may browse for the image file and then specify which numeric fields in the data source are used to determine the **X** and **Y** locations for the plots on the image. Additionally, you can specify how the values of these fields correspond to the width and height of the image:

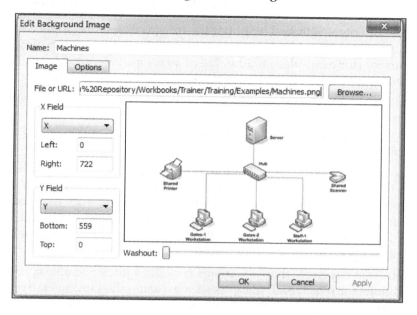

In the preceding screenshot, you can see the setup that allows us to create visualizations such as these:

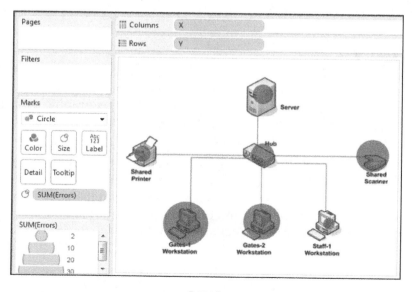

The **X** and **Y** fields have been used as continuous dimensions that result in a scatterplot. The axes have been hidden. A circle mark with the size determined by the sum of errors is plotted at each X and Y coordinate.

Here are some tips and tricks to remember when using background images:

- Make sure the width and height of the image correspond to the values of the fields used for X and Y. A 1:1 ratio between the fields and the dimension of the image (for example, a field value of 50 is equivalent to 50 pixels on the image) is recommended, but any consistent ratio can be used.

- In the preceding dataset, X and Y were already included in the data. You can use blending to supplement the data source with the X and Y values in the same way we supplemented data with latitudes and longitudes in an earlier example.

- Images are measured from 0 at the top to the height in pixels at the bottom. This is the reverse of the Y axis of a scatterplot, which starts from 0 at the bottom. If your image is upside down, check to make sure that you have specified a value for **Bottom** instead of **Top**. If the marks are plotted *upside down*, an easy fix is to right-click on the *y* axis of the scatterplot, select **Edit**, and then check **Reversed** under **Scale**.

- You can specify more than one image per data source and also specify the conditions for which each image is shown (using the **Options** tab of the **Edit Background Image** screen). This allows you to dynamically change images. This can be useful when multiple images apply to different segments of the data (for example, a different floor plan for each floor of the building), or when you want a way to dynamically change images based on a filter or an action.

Summary

We covered a wide variety of techniques in this chapter. We looked at sheet swapping, dynamic dashboards, using sets to answer complex questions, some advanced mapping techniques, including supplementing geographic data, adjusting map options, and using custom images. Finally, we looked at how to use background images to create unique and useful visualizations.

There is no way to cover every possible technique. Instead, the idea is to demonstrate some of what can be accomplished using a few advanced techniques. The examples in this chapter are built on the foundations laid in the rest of the book. From here, you will be able to creatively modify and combine techniques in new and innovative ways to solve problems and achieve incredible results.

11
Sharing Your Data Story

Throughout this book, we focused on Tableau Desktop and learned how to visually explore and communicate data with visualizations and dashboards. Once you've made discoveries, designed insightful visualizations, and built stunning dashboards, you're ready to share your data stories.

Tableau enables you to share your work using a variety of methods. In this chapter, we'll take a look at the various ways to share visualizations and dashboards, along with what to consider when deciding how you will share.

Specifically, we'll take a look at the following topics:

- Presenting, printing, and exporting
- Sharing with Tableau Desktop and Tableau Reader
- Sharing with Tableau Server, Tableau Online, and Tableau Public
- Options for presenting your data story
- Additional distribution options with Tableau Server

Presenting, printing, and exporting

Tableau is primarily designed to build richly interactive visualizations and dashboards for consumption on a screen. Often, you will expect users to interact with your dashboards and visualizations. However, there are good options for presenting, printing, and exporting in a variety of formats.

Presenting

Tableau Desktop and Reader features **Presentation Mode**, which is available from the **Window** menu or by pressing *F7* or using the option on the toolbar. This mode removes all authoring controls and displays only the view and navigation tabs in fullscreen. Press *F7* or *Esc* to exit the fullscreen display.

When used with effective dashboards and stories, Presentation Mode is an effective way to walk your audience through the data story.

If you save a workbook by pressing *Ctrl + S* while in Presentation Mode, the workbook will be opened in Presentation Mode by default.

Printing

Tableau enables printing for individual visualizations, dashboards, and stories. From the **File** menu, you can select **Print** to print the currently active sheet in the workbook to the printer, or select the **Print to PDF** option to export to a PDF. Either option allows you to export the active sheet, selected sheets, or the entire workbook to a PDF. To select multiple sheets, hold the *Ctrl* key and click on individual tabs.

When printing, you also have the option to select **Show Selections**. When this option is checked, marks that have been interactively selected or highlighted in a view or dashboard will be printed as selected. Otherwise, marks will be printed as though no selections have been made. The map in the following dashboard has marks for the eastern half of the United States selected:

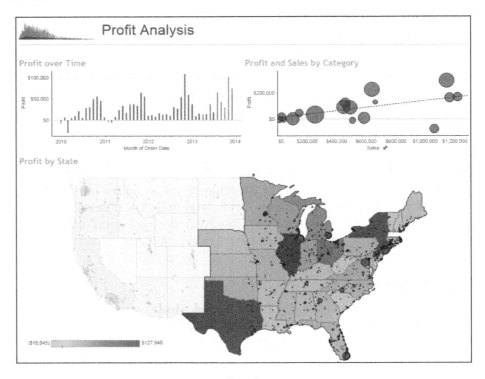

Here are some considerations, tips, and suggestions for printing:

- If a dashboard is being designed for print, select a predefined paper size as the fixed size for the dashboard or use a custom size that matches the same aspect ratio.

- Use the **Page Setup** screen (available from the **File** menu) to define specific print options, such as what elements (legends, title, and caption) will be included, the layout (including margins and centering), and how the view or dashboard should be scaled to match the paper size. The **Page Setup** options are specific to each view. Duplicating or copying a sheet will include any changes to the **Page Setup** settings:

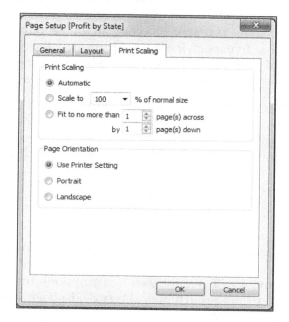

- Fields used in the **Pages** shelf will define page breaks in printing. The number of pages defined by the **Pages** shelf is not necessarily equivalent to the number of printed pages. This is because a single page defined by the **Pages** shelf might require more than one printed page.

- Each story point in a story will be printed on a new page.

- Printing the entire workbook can be an effective way to generate a single PDF document for distribution. Each visible sheet will be included in the PDF in the order of the tabs, left-to-right. You may hide sheets to prevent inclusion in the PDF or reorder sheets to adjust the order of the resulting document. Also, consider creating dashboards with images and text for title pages, table of contents, and commentary.

Sheets may be hidden if they are views that are used in one or more dashboards or if they are dashboards used in one or more stories. To hide a view, right-click on the tab or thumbnail on the bottom strip or in the left-hand pane of the dashboard or story workspace, and select **Hide Sheet**. To show a sheet, locate it in the left-hand pane of the dashboard or story workspace, right-click on it, and uncheck **Hide Sheet**.

Exporting

Tableau also makes it easy to export images of views, dashboards, and stories for use in documents, PowerPoint, and even books such as this one. Images may be exported as .png, .emf, .jpg, or .bmp. You may also copy an image to the clipboard to paste it into other applications. To export or copy an image, use the menu options for **Worksheet**, **Dashboard**, or **Story**. If you are viewing a dashboard, you can select an individual view, and use the worksheet menu to export an image of that single view.

We'll take a look at using Tableau Server, Tableau Online, and Tableau Public in detail shortly. For now, let's consider some of the available exporting features. When interacting with a web view, you will see a toolbar unless you don't have the required permissions, or the toolbar has been specifically disabled by a Tableau Server administrator:

The first button in the toolbar allows you to export an image, data, crosstab (Excel), or PDF. Images are exported as .png and render the dashboard in its current state. Exporting a PDF document will give the user many options, including layout, scaling, and whether to print the current dashboard, all sheets in the workbook, or all sheets in the current dashboard.

Sharing a workbook with users of Tableau Desktop or Tableau Reader

Sharing a workbook with other Tableau Desktop users is fairly straightforward, but there are a few things to consider.

One of the major considerations is whether you will be sharing a packaged workbook (.twbx) or an unpackaged workbook (.twb). Packaged workbooks are single files that contain the workbook (.twb), extracts (.tde), file-based data sources that have not been extracted (.xls, .xlsx, .txt, .cub, .mdb, and so on), custom images, and various other related files.

To share a workbook with users of Tableau Desktop:

- You may share either a packaged (.twbx) or unpackaged (.twb) workbook by simply sharing the file with another user who has the same or newer version of Tableau Desktop.

> Workbook files will be updated when saved in a newer version of Tableau Desktop. You cannot open a workbook saved with a newer version of Tableau in an older version. You will be prompted about updates when you first open the workbook and again when you attempt to save it.

- If you share an unpackaged (.twb) workbook, then anyone else using it must be able to access any data sources, and any referenced images must be visible to the user in the same directory where the original files were referenced. For example, if the workbook uses a **Live** connection to an Excel (.xlsx) file on a network path and includes images on a dashboard located in C:\Images, then all users must be able to access the Excel file on the network path and have a local C:\Images directory with image files of the same name.

- Similarly, if you share a packaged workbook (.twbx) that uses **Live** connections, anyone using the workbook must be able to access the live connection data source and have appropriate permissions.

Tableau Reader is a free product provided by Tableau Software that allows users to interact with visualizations, dashboards, and stories created in Tableau Desktop. Unlike Tableau Desktop, it does not allow authoring of visualizations or dashboards. However, all the interactivities, such as filtering, drill-down, actions, and highlighting, are available to the end user.

> Think of Tableau Reader as being similar to many PDF readers that allow you to read and navigate through the document but do not allow authoring.

To share a workbook with users of Tableau Reader:

- You may only share packaged (.twbx) workbooks.

- Furthermore, the packaged workbook may not contain **Live** connections to server or cloud-based data sources. These connections must be extracted.

Be certain to take into consideration security concerns when sharing packaged workbooks (.twbx). Since packaged workbooks most often contain the data, you must be certain that the data is not sensitive. Even if the data is not shown in any view or dashboard, it is still accessible in the packaged extract (.tde) or file-based data source.

Sharing data with users of Tableau Server, Tableau Online, and Tableau Public

Tableau Server, Tableau Online, and Tableau Public are all variations on the same concept: hosting visualizations and dashboards on a server and allowing users to access them via a web browser.

The following table provides some of the similarities and differences between the products, but as details may change, please consult with a Tableau representative prior to making any purchasing decisions:

Product	Tableau Server	Tableau Online	Tableau Public
Description	A server application installed on one or more server machines that host views and dashboards created with Tableau Desktop.	A cloud-based service maintained by Tableau Software that hosts views and dashboards created with Tableau Desktop.	A cloud-based service maintained by Tableau Software that hosts views and dashboards created with Tableau Desktop or the free Tableau Public client.
Licensing	It has a named user (a set number of users) or core (unlimited users on a set number of cores).	Named user.	Free.

Product	Tableau Server	Tableau Online	Tableau Public
Administration	This is fully maintained, managed, and administered by the individual or organization that purchased the license.	This is managed and maintained by Tableau Software with some options for project and user management by users.	This is managed and maintained by Tableau Software.
Authoring and publishing	Users of Tableau Desktop may author and publish views and dashboards to Tableau Server. Web authoring allows Tableau Server users the capability to edit and create visualizations in a browser.	Users of Tableau Desktop may author and publish views and dashboards to Tableau Online.	Users of Tableau Desktop or the free Tableau Public client can publish views and dashboards to Tableau Public.
Interaction	Licensed Tableau Server users, even those without Tableau Desktop, may interact with hosted views. Views may also be embedded in intranet sites, SharePoint, and custom portals.	Licensed Tableau Online users, even those without Tableau Desktop, may interact with hosted views. Views may also be embedded in intranet sites, SharePoint, and custom portals.	Anyone may interact with hosted views. Views may be embedded in public websites and blogs.
Limitations	None.	Most data sources must be extracted before workbooks can be published. Most non-cloud-based data sources must have extracts refreshed using Tableau Desktop on a local machine (though it can be scripted for automation).	All the data must be extracted and each data source is limited to 10 million rows.

Product	Tableau Server	Tableau Online	Tableau Public
Security	The Tableau Server administrator may create sites, projects, and users and adjust permissions for each. Access to the underlying data can be restricted and downloading of the workbook or data can be restricted.	The Tableau Server administrator may create projects and users and adjust permissions for each. Access to the underlying data can be restricted and downloading of the workbook or data can be restricted.	By default, anyone may download and view data; however, access may be restricted.
Good uses	This is used for internal dashboards and analytics and/or use across departments/divisions/clients via multitenant sites.	This is used for internal dashboards and analytics. It is also used for sharing and collaborating with remote users.	This is used for sharing visualizations and dashboards using embedded views on public-facing websites or blogs.

Publishing to Tableau Public

You may open and save workbooks to Tableau Public using either Tableau Desktop or the free Tableau Public client application. Please keep the following points in mind:

- In order to use Tableau Public, you will need to register an account
- With Tableau Desktop, you may save and open workbooks to and from Tableau Public using the **Server** menu, and selecting options under **Tableau Public**
- With the free Tableau Public client, you may only save workbooks to and from the Web
- Selecting the option to **Manage Workbooks** will open a browser, so you can log in to your Tableau Public account and manage all your workbooks online
- Workbooks saved to Tableau Public may contain any number of data source connections, but they must all be extracted and must not contain more than 10 million rows of extracted data

When you publish a workbook to Tableau Public, you will be shown a preview and given a link where you can view it online in a web browser. The following story is hosted on Tableau Public (you can view it at `http://www.tableausoftware.com/public/gallery/second-punic-war`):

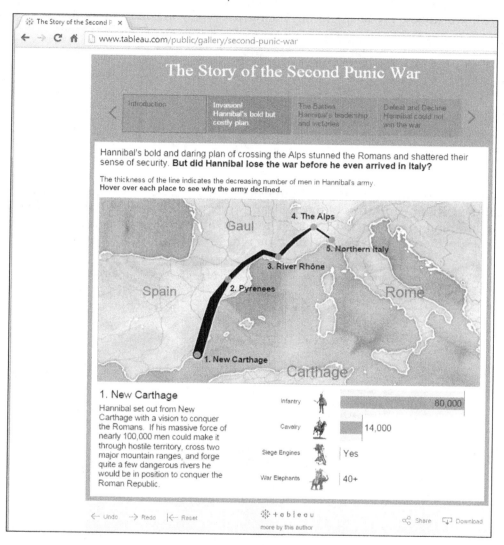

You'll notice the toolbar at the bottom that gives you the ability to **Undo** and **Redo** various interactions or **Reset** the view to its original state. Additionally, there are links to learn more about Tableau and the individual author of the visualization. Finally, there are options for downloading the workbook, PDF, image, or data, and for sharing the workbook via link, embed code, or social media.

Publishing to Tableau Server and Tableau Online

To publish to Tableau Server or Tableau Online, from the menu navigate to **Server | Publish Workbook**. If you are not signed into a server, you will be prompted to sign in:

You must have a user account with publish permissions for one or more projects. Enter the URL or IP address of the Tableau Server or the Tableau Online URL, your username, and password. Once signed in, you will be prompted to select a site if you have access to more than one. Finally, you will see the publish screen:

Here, you will be able to select the project to which you wish to publish and name your workbook. If a workbook has already been published with the same name to the selected project, you will be prompted to overwrite it. You may use **Add Tags** to make searching for and finding your workbook easier.

User/Group permissions related to who has permissions to view, interact with, and alter your workbook will start with the default settings for the project based on any setup done by the Tableau Server administrator. As the publisher, you may customize these permissions.

You may also specify the views to share using the **Views to Share** option. Any sheets you check will be included in the published workbook. Any sheets you uncheck will not.

You also have the option to select **Show Sheets as Tabs**. When this is checked, users on Tableau Server will be able to navigate between sheets using tabs similar to those shown at the bottom of Tableau Desktop. This option must be checked if you plan to have actions that navigate between views. **Show Selections** indicates that you may wish to have any active selections of marks retained in the published views.

The **Scheduling & Authentication...** button opens a screen that allows you to accomplish two important tasks:

- For each data connection used in the workbook, you may determine how the database connections are authenticated. The options will depend on the data source as well as the configuration of Tableau Server. Various options include embedding a password, impersonating a user, or prompting a Tableau Server user for credentials.

- You may specify a schedule for Tableau Server to run refreshes of any data extracts.

Any live connections or extracted connections that will be refreshed on the server must define connections that work from the server. This means that all applicable database drivers must be installed on the server; all networks, Internet connections, and ports required for accessing database servers and cloud-based data must be open.

Additionally, any external files referenced by a workbook (for example, image files and nonextracted file-based data sources) that were not included when the workbook was published must be referenced using a location that is accessible by Tableau Server (for example, a network path with security settings allowing the Tableau Server process to have read access).

Interacting with Tableau Server

After a workbook is published to Tableau Server, other users will be able to view and interact with the visualizations and dashboards. Once logged into Tableau Server, they will be able to browse content for which they have appropriate permissions. These users will be able to use any features built into the dashboards, such as quick filters, parameters, actions, or drilldowns. Everything is rendered as HTML 5, so the only requirement for the user to view and interact with views and dashboards is a web browser.

[The **Tableau Mobile** app, which is available for iPad and Android devices, can enhance the experience for mobile users.]

Here is an example of a dashboard published to Tableau Server:

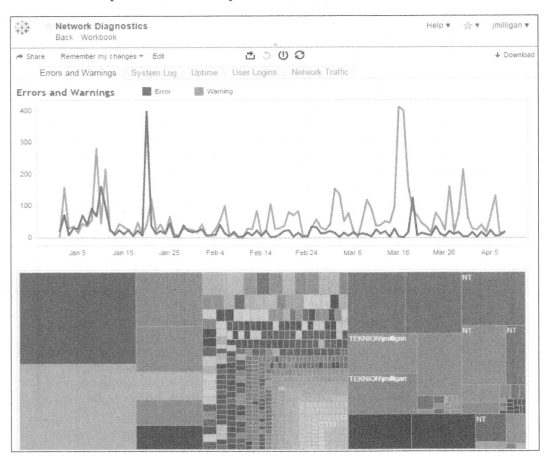

For the most part, interacting with a workbook on Server or Online is very similar to interacting with a workbook in Tableau Desktop or Reader. Quick filters, parameters, actions, and tooltips all look and behave similarly.

You will find some additional features:

- The top menu gives you various options related to managing and navigating to Tableau Server.

- Below that, you'll find a breadcrumb trail giving the hierarchy of the current workbook and view.

- Beneath that, you'll find a link for options to share the workbook. These options include a URL that you can distribute to other licensed users as well as code for embedding the dashboard in a web page:

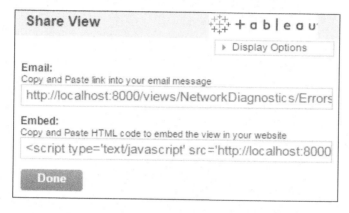

- The **Remember My Changes** link allows you to take and manage snapshots of the dashboard in its current state (your current selections, filters, and parameter selections). This link also allows you to navigate to previously saved dashboard snapshots:

- The **Edit** link opens a limited web authoring mode. The interface is very similar to Tableau Desktop. Web editing currently allows only a single view to be edited or created and not an entire dashboard. The Tableau Administrator can enable or disable web editing per user or group and also control permissions for saving edited views:

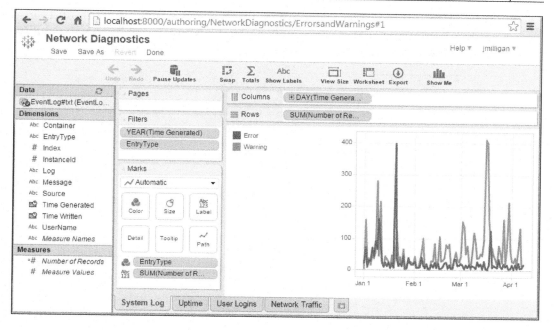

- The default toolbar contains four buttons: ☝ ↻ ⏻ ↺ . The first button allows you to export images, data, PDF, or Excel files based on the data. The Tableau Administrator may control permissions for these options. The second button reverts the view to the original state, including all filters, actions, and parameter values. The third button pauses the updates that allow you to make changes to multiple filters and parameter values without having to wait for updates until you are ready to proceed. The final button refreshes the view that clears the cache and requeries the data source to retrieve the latest data.

- The **Download** link allows the workbook to be downloaded so that Tableau Desktop and Tableau Reader users can interact with it locally. Tableau administrators can disable this option.

- Tabs (shown if the dashboard was published with the **Show Sheets as Tabs** option or if the option was subsequently enabled on Tableau Server) allow the user to navigate between visible sheets and dashboards in the workbook.

- Beneath the view or dashboard is a section for reviewing comments and adding comments or tags. This enables collaboration and enhances the ability to search for and find a specific view or dashboard on Tableau Server:

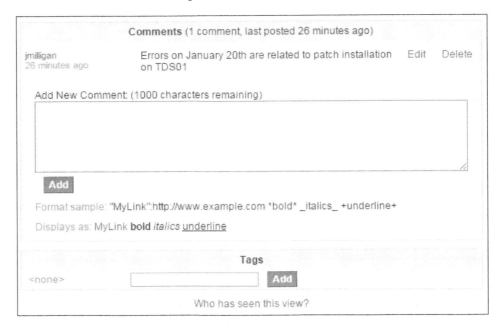

Additional distribution options using Tableau Server

Tableau Server allows several other options for sharing your views, dashboards, and data. Along with allowing users to sign in to Tableau Server, you might consider the following options:

- Dashboards and views can be embedded in websites, portals, and SharePoint. Single sign-on options exist to allow your website authentication to integrate seamlessly with Tableau Server.

- Tableau Server allows users to subscribe to views and dashboards and schedule e-mail delivery. The e-mail will contain an up-to-date image of the view and a link to the dashboard on Tableau Server.

- The **tabcmd** utility is provided with Tableau Server and may be installed on other machines. The utility provides you with the ability to automate many functions of Tableau Server, including export features, publishing, and user and security management. This opens up quite a few possibilities for automating delivery.

Summary

Tableau is an amazing platform for building useful and meaningful visualizations and dashboards based on your data. We considered how to connect to the data, write calculated fields, and design dashboards. In this chapter, we considered how to share the results with others.

You now have a solid foundation. At its core, Tableau is intuitive, transparent, and easy to use. As you dive deeper, the simplicity becomes increasingly beautiful as you discover new paradigms, solve complex problems, ask new questions, and find new answers in your data.

Index

S

scatterplot 107
scope
  about 153, 154
  options 154
  working with 155-157
security considerations, extracts 50
sets
  complex questions, answering 277-281
  leveraging, to answer complex
      questions 275-277
sheet
  selecting 268
  swapping 268-272
Show Me toolbar
  about 23
  example 25
  features 23, 24
stacked bars 91-94
star schema
  about 249, 250
  example 249
story points
  about 213
  example 213-218
surrogate key 250
symbol map, geographic
      visualizations 21, 22

T

Tableau
  about 1
  Access data, connecting to 2, 3
  dashboard 25
  data structure 245, 246
  data, visualizing 9
  distributions 236-238
  exporting 296
  forecasting 238-243
  formatting, working 175
  main workspace 4
  poorly shaped data, in
      visualizations 256-259
  presenting 293, 294
  printing 294, 295

  trends 222-225
  URL, for Viz of the Day 218
Tableau connections
  data structure 252-255
Tableau Data Extract (TDE) 40
Tableau Data Source (TDS) file 52
Tableau Desktop
  workbook, sharing with users 297
Tableau Online users
  publishing to 302, 303
Tableau Packaged Workbook (TWBX)
    file 49
Tableau paradigm
  about 31, 32
  example 33-36
Tableau Public
  publishing to 300, 301
Tableau Reader
  workbook, sharing with users 298
Tableau Server
  interacting with 304-308
  publishing to 302, 303
  used, for distribution options 308
Tableau Workbook (TWB) file 49
table calculations
  about 118, 148
  advanced 161
  aggregation 148
  creating 149, 150
  editing 149, 150
  filtering 148
  late filtering 149
  performance 149
  quick table calculations 150-153
table calculations, advanced
  Lookup and previous value 162, 163
  Metatable functions 161, 162
  Rank functions 164
  R script functions 165
  Running functions 163
  Total function 165
  Window functions 164
table calculations, examples
  last occurrence 169-171
  late filtering 168, 169
  Moving Average 165, 166
  ranking, within higher levels 167, 168

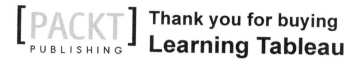

About Packt Publishing

Packt, pronounced 'packed', published its first book, *Mastering phpMyAdmin for Effective MySQL Management*, in April 2004, and subsequently continued to specialize in publishing highly focused books on specific technologies and solutions.

Our books and publications share the experiences of your fellow IT professionals in adapting and customizing today's systems, applications, and frameworks. Our solution-based books give you the knowledge and power to customize the software and technologies you're using to get the job done. Packt books are more specific and less general than the IT books you have seen in the past. Our unique business model allows us to bring you more focused information, giving you more of what you need to know, and less of what you don't.

Packt is a modern yet unique publishing company that focuses on producing quality, cutting-edge books for communities of developers, administrators, and newbies alike. For more information, please visit our website at www.packtpub.com.

Writing for Packt

We welcome all inquiries from people who are interested in authoring. Book proposals should be sent to author@packtpub.com. If your book idea is still at an early stage and you would like to discuss it first before writing a formal book proposal, then please contact us; one of our commissioning editors will get in touch with you.

We're not just looking for published authors; if you have strong technical skills but no writing experience, our experienced editors can help you develop a writing career, or simply get some additional reward for your expertise.

PUBLISHING

Building Interactive Dashboards with Tableau [Video]

ISBN: 978-1-78217-730-2   Duration: 04:31 hours

Create a variety of fully interactive and actionable Tableau dashboards that will inform and impress your audience!

1. Increase your value to an organization by turning existing data into valuable, engaging business intelligence.

2. Master the dashboard planning process by knowing which charts to use and how to create a cohesive flow for your audience.

3. Includes best practices and efficient techniques to walk you through the creation of five progressively engaging dashboards.

Tableau Data Visualization Cookbook

ISBN: 978-1-84968-978-6   Paperback: 172 pages

Over 70 recipes for creating visual stories with your data using Tableau

1. Quickly create impressive and effective graphics which would usually take hours in other tools.

2. Lots of illustrations to keep you on track.

3. Includes examples that apply to a general audience.

Please check **www.PacktPub.com** for information on our titles

Highcharts Cookbook

ISBN: 978-1-78355-968-8   Paperback: 332 pages

80 hands-on recipes to create, integrate, and extend dynamic and interactive charts in your web projects

1. Create amazing interactive charts that update in real time.

2. Make charts that work wherever you go: phone or desktop; online or offline.

3. Learn how to extend, and enhance Highcharts to design and develop charts easily.

4. Learn how you can integrate charts into just about any project for analyzing statistical data.

Mastering D3.js

ISBN: 978-1-78328-627-0   Paperback: 352 pages

Bring your data to life by creating and deploying complex data visualizations with D3.js

1. Create custom charts as reusable components to be integrated with existing projects.

2. Design data-driven applications with several charts interacting between them.

3. Create an analytics dashboard to display real-time data using Node and D3 with real world examples.

Please check **www.PacktPub.com** for information on our titles

CPSIA information can be obtained at www.ICGtesting.com
Printed in the USA
LVOW03s1531260415

436148LV00029B/832/P